Rapid Practical Designs
of Active Filters

Rapid Practical Designs
of Active Filters

DAVID E. JOHNSON

JOHN L. HILBURN

A Wiley-Interscience Publication

JOHN WILEY & SONS

New York / London / Sydney / Toronto

Library of Congress Cataloging in Publication Data:

Johnson, David E
  Rapid, practical designs of active filters.

  "A Wiley-Interscience publication."
  Includes bibliographical references and index.
  I.   Electric filters—Design and construction.
I.   Hilburn, John L., 1938–   joint author.   II.   Title.

TK7872.F5J64   621.3815'32   75–14074

ISBN 0-471-44304-2

Printed in the United States of America

10 9 8 7 6 5 4 3 2 1

Preface

This book presents simplified, rapid methods for obtaining a complete, practical filter design by inspection of a table. The book is intended for all filter designers, from the novice to the expert. The filter circuit elements used are integrated circuit (IC) operational amplifiers, resistances, and capacitances. All design tables are developed for standard, commonly available capacitor values.

From these tables the following types of filters may be constructed.
1. Low-pass (Butterworth or Chebyshev of second through eighth orders).
2. High-pass (Butterworth or Chebyshev of second through eighth orders).
3. Bandpass (Butterworth or Chebyshev of second through eighth orders).
4. Band-reject or notch (Butterworth or Chebyshev of second or fourth orders).
5. Phase-shift or all-pass (second order).
6. Constant-time-delay or Bessel (second, third, or fourth orders).
7. All-pass constant-time-delay (second order).

In the Chebyshev cases, ripple widths of 0.1, 0.5, 1, 2, and 3 dB are given for almost all filter designs.

Each filter type is discussed in a separate chapter. At the end of each chapter the design procedure is summarized and the appropriate tables are presented. Practical design suggestions are given for each circuit considered. The most popular filter designs, such as VCVS, infinite-gain multiple-feedback, and biquad, have been included in the chapter of each type. In addition, several multiple-feedback filters of our own design, which have superior performance

v

characteristics, have been included for the low-pass, high-pass, bandpass, and delay cases. A new type of filter, which exhibits both Bessel and all-pass features, is discussed in Chapter 6.

Numerous detailed examples are given for most filter types considered and actual photographs of the results are included. For instance, a detailed design of a low-pass VCVS filter is given in Sec. 2.5, which may be used as a general guideline. However, it is not necessary to read the chapters in order to use the book, since all the necessary information is presented on the summary pages at the end of each chapter.

This is the second book on this subject that we have coauthored, and since equal contributions have been made by each author on each book, the order of our names has been reversed on this book.

DAVID E. JOHNSON
JOHN L. HILBURN

Baton Rouge, Louisiana
February 1975

Contents

Rapid Practical Designs
of Active Filters

1
Introduction

1.1 Elements of Filter Theory

An electric filter is a device that passes signals of certain frequencies and rejects or attenuates those of other frequencies. The frequencies that pass constitute the *passbands* and those that are attenuated make up the *stopbands* of the filter. The location in the frequency domain of the passbands and stopbands is used to classify the filter as low-pass, high-pass, bandpass, band-reject, or all-pass. These types, as well as a constant-time-delay type, will be discussed in the remainder of the book, with a chapter devoted to each.

The performance of a filter may be measured by its *amplitude response*, which is a plot of the amplitude $|H(j\omega)|$ of its transfer function $H(s)$ versus frequency ω (radians/sec) or f (Hz), where $\omega = 2\pi f$. In all cases, we shall take $H(s) = V_2(s)/V_1(s)$, where V_2 and V_1 are the output and input voltages, respectively, of the filter. Examples of ideal amplitude responses and physically realizable approximations to the ideal cases will be presented for each filter type in the appropriate chapter.

In addition to the amplitude response, another important characteristic of a filter is its *phase response*, $\phi(\omega)$, plotted versus ω, where $\phi(\omega)$ is defined by

$$H(j\omega) = |H(j\omega)|e^{j\phi(\omega)} \tag{1.1}$$

In some filter designs, such as the all-pass and Bessel filters considered in Chapter 6, the phase response is the characteristic of interest.

In an earlier work [1]*, we considered rapid design techniques for lower-order active filters, principally second and fourth orders. (The term "order"

* References thus cited are listed at the end of the book.

1

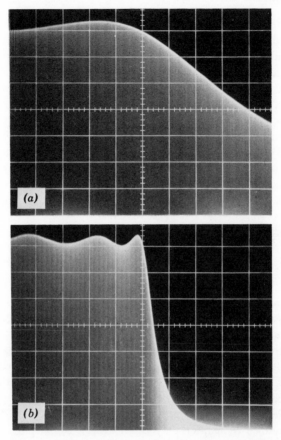

Figure 1-1. (*a*) A second-order and (*b*) a sixth-order low-pass filter amplitude response

will be defined later for each filter type, where it will be seen that the higher the order the more sharply separated are the passbands and stopbands and hence the better the filter.) An illustration of the merits of a higher-order filter versus a lower-order filter is given in Fig. 1-1. Part (*a*) shows the amplitude reponse of a second-order low-pass filter and (*b*) shows the response of the same type low-pass filter of sixth order. These are responses of actual circuits whose construction will be described in Chapter 2.

Passive filters are constructed with inductors, capacitors, and resistors, and are very practical for certain frequency ranges. For lower frequency ranges, however, inductors, because of their size and practical performance limitations, are undesirable. Consequently, designers for many years have sought to replace inductors by active devices that simulate the effect of inductors. In recent years, the trend toward active devices has accelerated with the advances in miniaturization which have made the active elements available at prices competitive with, and in most cases cheaper than, those of inductors.

There are many ways of obtaining filters using active devices in lieu of inductors. (See, e.g., references [2] through [8].) The active device we use is the integrated circuit (IC) operational amplifier, which is briefly described in the next section. Most of the designs we present are those currently being used in constructing active filters. Also, for most filter types we present a new circuit of our own design. Tables are presented for each filter type and, depending on the specifications, the designer may simply choose the appropriate table and read off the circuit element values. For the reader who is interested in the theoretical details, there is a chapter of background material with numerous references provided for each filter type. However, to use the design procedure to construct a filter, one needs only to refer to the summary sheet preceding the tables at the end of each chapter.

1.2 The Operational Amplifier

The symbol that we use for our basic active element, the operational amplifier (op-amp), is shown in Fig. 1-2. The op-amp is a multiterminal device, but for simplicity we have shown only three terminals, the inverting input terminal $(-)$, the noninverting input terminal $(+)$, and the output terminal. The output voltage depends upon the difference between the voltages at the input terminals, and hence the op-amp shown in Fig. 1-2 is often referred to as a differential op-amp. The purposes of the terminals that are not shown are specified by the manufacturer and include, in general, power supply connections, frequency compensation terminals, and offset null terminals.

The equations we have derived in the following chapters are obtained assuming zero voltage between and zero current into the two input terminals of the op-amp. This is true of the ideal op-amp and is closely approximated by practical op-amps, if used according to the manufacturer's specifications.

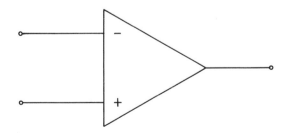

Figure 1-2. A differential op-amp.

Other features of ideal op-amps, approximated by practical op-amps, are their infinite input and zero output impedance characteristics. These allow one to cascade two or more circuits without appreciably affecting their individual operations.

Numerous publications are available describing in detail the characteristics and uses of commercially available op-amps. (See, e.g., references [2], [4] and [9] through [18].) In addition, most manufacturers publish detailed catalogs containing information on their specific op-amps. An extensive list of manufacturers is given in [9]. Some well-known manufacturers include Burr-Brown Research Corp., Fairchild Semiconductor, Motorola, National Semiconductor, RCA, Signetics Corp., and Texas Instruments.

In general, for stable operation, IC op-amps require frequency compensation. Some are internally compensated, such as the 741, 536, 107, 5556, 740, and 747 (dual 741). Different representations are used to identify the manufacturer. For example, the μA741, AD741, MC1741, LM741, RC741, SN72741, CA3056A, and so forth are all type 741 op-amps. Other op-amps require external compensation, specified by the manufacturer, but generally are useful for much higher frequencies and gains. Some examples of these are the types 709, 748, 101, and 531.

For best results in the circuit configurations given in the following chapters, the designer should use op-amps that perform adequately for the gains and frequency ranges of interest. For example, the open-loop gains as specified by the manufacturer should be at least 50 times the filter gain [11]. Other suggestions will be made on the summary sheets at the end of each chapter.

1.3   Resistors and Capacitors

Three types of resistors in common use are the carbon composition, metal-film, and wire-wound resistors. Of the three, the carbon composition is the most economical and most used in noncritical filter design. This is particularly true if the filter is used at room temperature. In our examples, the second-, third-, and fourth-order filters were constructed with 5% tolerance carbon composition resistors. For high-performance applications, or in instances where temperature is important, one should use either metal-film or wire-wound resistors. For fifth- and sixth-order filters, the design is more critical, and elements with approximately 2% tolerances are required. Seventh- and eighth-order designs are even more critical, and elements of 1% tolerances should be used. In our examples of fifth- through eighth-order filters, both wire-wound and metal-film resistors of 1% tolerance were used.

In the case of capacitors, the ceramic disk is a very common and economical type. However, it should be used in only the most noncritical applications. A more acceptable common type is the Mylar capacitor, which is the type we used in most of our filter examples. For critical applications and high performance, polystyrene and Teflon capacitors are good choices in most cases. For a good discussion of resistors and capacitors, the reader is referred to reference [2], pp. 317–319.

As a final note, construction of higher-order filters, such as seventh- and eighth-order, should be attempted only by the experienced designer with the equipment for measuring and testing his circuit elements.

2

Low-Pass Filters

2.1 General Theory

A low-pass filter is a device that passes signals of low frequencies and attenuates or rejects those of high frequencies. An ideal low-pass amplitude response, represented by the broken line, is shown in Fig. 2-1 with a realizable approximation to the ideal, represented by the solid line. The passband, $0 < \omega < \omega_c$, and the stopband $\omega > \omega_c$ are clearly indicated in the ideal case, but in the actual case, the *cutoff frequency* ω_c (or, in Hz, $f_c = \omega_c/2\pi$) must be defined. The usual definition of ω_c is the point at which $|H(j\omega)|$ is $1/\sqrt{2} = 0.707$ times its maximum value, shown in Fig. 2-1 as A.

Some authors consider the stopband to begin at a point $\omega_1 > \omega_c$ where $|H(j\omega_1)|$ is considerably below $A/\sqrt{2}$ and define a *transition band* $\omega_c < \omega$

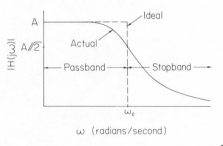

Figure 2-1. Low-pass amplitude responses.

6

$< \omega_1$. For our purposes this refinement is not necessary, and we shall simply refer to the stopband as the interval $\omega > \omega_c$.

The amplitude response may also be plotted in decibels (dB), denoted by α and defined by

$$\alpha = 20 \log_{10} |H(j\omega)| \qquad (2.1)$$

Thus, at cutoff, α is $20 \log_{10} \sqrt{2} = 3$ dB below its maximum value.

The ideal low-pass amplitude response cannot be physically realized, of course, but approximations to the ideal low-pass filter may be achieved by the transfer function

$$\frac{V_2}{V_1} = \frac{Gb_0}{s^n + b_{n-1}s^{n-1} + \cdots + b_1 s + b_0} \qquad (2.2)$$

where $G, b_0, b_1, \ldots, b_{n-1}$ are appropriately chosen constants. Equation (2.2) represents an nth order, *all-pole* approximation, so called because its denominator polynomial is of nth degree and its numerator is a constant (thus it has no finite zeros, only finite poles).* The *gain* of the low-pass filter is the value of its transfer function at $s = 0$ and in the case of Eq. (2.2), the gain is evidently G.

We shall restrict ourselves to all-pole functions so that Eq. (2.2) shall represent the general case for various values of n. For types other than all-pole approximations, as well as for a thorough discussion of low-pass filter theory, the reader is referred to such standard filter and circuit theory books as [19] through [22], and such recent books as [23] and [24].

There are a number of types of low-pass filters, but probably the most commonly used are the *Butterworth* and *Chebyshev* types. Their transfer functions are of the type of Eq. (2.2) and differ only in the choice of constants $b_0, b_1, \ldots, b_{n-1}$. In this chapter we shall discuss briefly the properties of Butterworth and Chebyshev low-pass filters and present simple and rapid techniques for their design.

2.2   Low-Pass Butterworth Filters

A filter that approximates the ideal low-pass filter with a so-called *maximally flat* passband characteristic is the *Butterworth* filter [21], [22]. Its amplitude response is given by

$$|H(j\omega)| = \frac{G}{\sqrt{1 + (\omega/\omega_c)^{2n}}}; \qquad n = 1, 2, 3, \ldots \qquad (2.3)$$

---

* Note: A zero is a value of s for which the function is zero and a pole is a value of s for which the function becomes infinite.

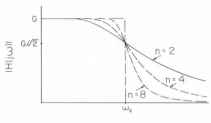

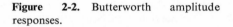

Figure   2-2. Butterworth   amplitude responses.

where n is the order of the filter. As n increases, the response more nearly approximates the ideal, as may be seen in Fig. 2-2, where numerous cases are shown.

The Butterworth filter has excellent amplitude characteristics near $\omega = 0$, but its cutoff characteristics (near $\omega = \omega_c$) are relatively poor, as is its rate of attenuation into the stopband. A filter that sacrifices some of the desirable low-frequency qualities of the Butterworth, but which is far superior in all other amplitude characteristics is the Chebyshev filter, to be described in the next section. As an example, for relatively high frequencies, the attenuation rate of the Butterworth response is approximately $-20n$ dB/decade, which is distinctly inferior to that of the Chebyshev filter [21]. (A decade is the frequency interval between two frequencies, one being 10 times the other).

The ideal phase response is linear with slope $-\tau$, where τ is the *time-delay* to be discussed in Chapter 6. From the standpoint of phase response, the Butterworth filter is superior to the Chebyshev filter; that is, its phase response is more nearly linear over the passband [21]. As a general rule, the better the amplitude response of a filter, the poorer its phase response, and vice versa. Phase responses of the Butterworth cases given in Fig. 2-2 are shown in Fig. 2-3.

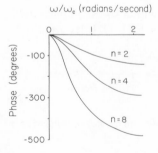

Figure 2-3. Butterworth phase responses.

2.3 Low-Pass Chebyshev Filters

The low-pass Chebyshev filter has amplitude response [22] given by

$$|H(j\omega)| = \frac{K_1}{\sqrt{1 + \epsilon^2 C_n^2(\omega/\omega_c)}}; \qquad n = 1, 2, \ldots \qquad (2.4)$$

where ϵ and K_1 are constants and C_n is the Chebyshev polynomial of the first kind of degree n. Examples for various values of n and a single value of ϵ are shown in Fig. 2-4, where it may be seen that there are ripples of equal width in the passband. As n increases, the number of ripples increases and the attenuation in the stopband increases. This yields, for a fixed deviation in the passband, a better approximation with increasing n to the ideal response.

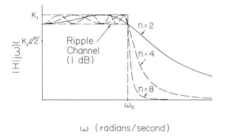

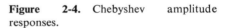

ω (radians/second)

Figure   2-4. Chebyshev   amplitude responses.

   The ripple width, given for $K_1 = 1$ by $RW = 1 - 1/\sqrt{1 + \epsilon^2}$, and defined in decibels by

$$RW_{dB} = 20 \log \sqrt{1 + \epsilon^2} = 10 \log (1 + \epsilon^2)$$

is determined by selecting ϵ, and may be used to characterize the filter. For example, a 1 dB Chebyshev low-pass filter is one whose response has a pass-band ripple width, expressed in dB, of 1. Referring to Fig. 2-4, it is clear that ω_c, as defined, is the end of the ripple channel and not the conventional 3 dB cutoff point, except in the case of a 3 dB filter. In our discussion of the Chebyshev filter, we shall refer to ω_c as the cutoff point meaning the terminal frequency of the ripple channel. To relate this value of ω_c to the conventional 3 dB cutoff frequency, one may consult Table 2-1, for various values of n and ripple widths (decibels).

   In applications where passband ripples are undesirable, the Butterworth filter is evidently preferable to the Chebyshev. However, for a fixed n and given allowable deviation in the passband, the Chebyshev filter is the best of all the all-pole filters in that it has the smallest possible transition interval from the passband to some specified attenuation in the stopband [25]. For example, the attenuation in decibels of the Chebyshev amplitude response is

Table 2-1.   Ratio of Conventional Cutoff $f_{3\,\mathrm{dB}}$ to Channel Terminal f_c for Low-Pass Chebyshev Filters

	$f_{3\,\mathrm{dB}}/f_c$						
dB \ n	2	3	4	5	6	7	8 ·
0.1	1.943	1.389	1.213	1.135	1.093	1.068	1.052
0.5	1.390	1.168	1.093	1.059	1.041	1.030	1.023
1	1.218	1.095	1.053	1.034	1.023	1.017	1.013
2	1.074	1.033	1.018	1.012	1.008	1.006	1.005
3	1.000	1.000	1.000	1.000	1.000	1.000	1.000

approximately $3(n - 1) + 20 \log \epsilon$ below that of the Butterworth in the stop-band [21].

As was discussed in the previous section, the phase response of the Chebyshev filter is inferior to that of the Butterworth filter. Higher-order Chebyshev filters have better amplitude responses and therefore poorer phase responses. Several Chebyshev phase responses are shown in Fig. 2-5, which illustrates this property.

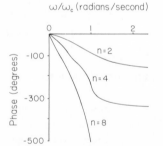

Figure 2-5.   Chebyshev phase responses.

2.4   Design Tables

The coefficients $b_0, b_1, b_2, \ldots$ of Eq. (2.2) may be calculated using Eq. (2.3) in the Butterworth case and Eq. (2.4) in the Chebyshev case, and are well-documented (see, e.g., reference [22]). For the convenience of the reader, the coefficients of the general nth degree polynomial, as well as those of its second-order factors, are given in Appendices A and B for the Butterworth and Chebyshev functions for $n = 1, 2, \ldots, 8$. The cutoff frequency ω_c has been normalized to 1 rad/sec in every case.

To obtain filters with a given cutoff frequency f_c (hertz), we may use the normalized coefficients and scale the resulting network. There are many ways to do this but we have chosen the following method. We design the normalized network with normalized capacitances, such as $C = 1$ F. Then we replace the normalized 1 F capacitances by a practical standard value C and multiply each normalized resistance by the factor

$$k = \frac{1}{2\pi f_c C} \tag{2.5}$$

Normalized capacitors other than C are denormalized by multiplying their values by C.

For the filter circuits discussed in the remainder of the book, the scaling has been accomplished by means of design tables, assembled at the end of each chapter. The scale factor k has been divided into two factors, one of which is incorporated into the tables. The other is the K *parameter* given by

$$K = \frac{100}{f_c C'} \tag{2.6}$$

where C' is the value of C in microfarads. That is, if C selected is 0.01 μF, then $C' = 0.01$, and so forth. The tables are constructed so as to yield resistance values in kilohms for a K parameter of 1. The scaling is completed by multiplying the resistances of the tables by K given in Eq. (2.6).

Alternatively, for the convenience of the designer, graphs are included in every chapter for use in lieu of Eq. (2.6) in finding K for a given f_c and selected C'. For higher-order filters, where more accuracy is required, it is probably better to use Eq. (2.6) rather than the graphs to find K.

2.5   VCVS Low-Pass Filters

There are many ways of constructing low-pass filters using operational amplifiers, resistors, and capacitors. In the remainder of this chapter we consider some of the more common methods currently being used, as well as one new method of our own design. One general-purpose circuit which is widely used is that of Sallen and Key [26]. We refer to the Sallen and Key circuit as a VCVS low-pass type because it uses an op-amp and two resistors connected so as to constitute a voltage-controlled-voltage-source (VCVS).

In the case of second-order low-pass filters, the transfer function (2.2) becomes

$$\frac{V_2}{V_1} = \frac{Gb_0}{s^2 + b_1 s + b_0} \tag{2.7}$$

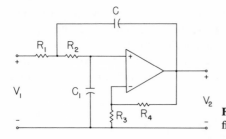

Figure 2-6. A second-order VCVS low-pass filter.

The Sallen and Key circuit which realizes this function is shown in Fig. 2-6, where the op-amp and resistors R_3 and R_4 constitute the VCVS. Higher even-order low-pass filters may be obtained by cascading two or more second-order filters or stages. In this case the transfer function must be factored into a product of second-order functions such as (2.7). Each stage then may be realized by a circuit like that of Fig. 2-6. (See, e.g., reference [27].)

Analysis of Fig. 2-6 shows that it achieves Eq. (2.7) with

$$b_0 = \frac{1}{R_1 R_2 C C_1}$$

$$b_1 = \frac{1}{R_2 C_1}(1 - \mu) + \frac{1}{R_1 C} + \frac{1}{R_2 C} \tag{2.8}$$

$$G = \mu = 1 + \frac{R_4}{R_3}$$

where μ is the gain of the VCVS and also the gain of the filter.

To design a low-pass VCVS second-order filter or a second-order stage of a higher-order filter, we use Fig. 2-6 and determine practical values of the capacitances C and C_1 and the resistances R_1, R_2, R_3, and R_4 so that Eqs. (2.8) are satisfied for a given gain G and coefficients b_0 and b_1.

To facilitate the design procedure we have constructed a number of tables that may be used to obtain the capacitances and resistances, as follows. Given a desired cutoff frequency f_c, gain G, and type (Butterworth or Chebyshev, etc.), we select a standard value of capacitance C and determine a K parameter from Eq. (2.6) or from one of Figs. 2-16a through 2-16c. Using this value of K, we then select the appropriate one of Tables 2-2 through 2-25 to find C_1 and the resistances for $K = 1$. We then multiply the resistances by K to obtain the required resistances for the filter design.

As an example, suppose we desire a second-order VCVS 0.5 dB Chebyshev low-pass filter with a gain of $G = 2$ and a cutoff frequency (terminal frequency of the ripple channel) $f_c = 2000$ Hz. Selecting a capacitor C of 0.01 μF we have from Fig. 2-16b a K parameter of 5. Alternatively, the K parameter may also be found from Eq. (2.6). From Table 2-4, we find C_1 and the resistances for a K parameter of 1. Hence, we must multiply the resistances by 5, resulting in the network elements $R_1 = 5.580$, $R_2 = 7.485$, $R_3 = R_4 = 26.130$ (all kilohms), and $C_1 = C = 0.01 \mu F$. The filter is then constructed in accordance with Fig. 2-6. The response in this case was given earlier in Fig. 1-1a. Numerous examples of second-order filters were constructed and illustrated in [1] with actual circuit element values and responses shown. The resistances used were as close as possible to the calculated values.

The VCVS filter is one of the more popular low-pass filters having *non-inverting* (positive) gain. Some of its advantages are that it has a minimal number of network elements, a relative ease of adjustment of characteristics, a low output impedance, a low spread of element values, and is capable of relatively high gains [28]. A definite advantage is the ability to precisely set the gain of the filter by setting the gain of the VCVS by means of a potentiometer. This technique is described in the complete design procedure for the construction of the VCVS filter, given in Sec. 2.10.

2.6 Infinite-Gain-Multiple-Feedback Low-Pass Filters

A circuit that realizes Eq. (2.7) with one less resistor than the Sallen and Key network is the multiple-feedback circuit [2] of Fig. 2-7. We call this filter an *infinite-gain multiple-feedback* (infinite-gain MFB) low-pass filter because of its two feedback paths through C_1 and R_2, and because the op-amp is serving as an infinite gain device rather than a finite gain VCVS, as in the Sallen and Key circuit. In addition, the adjective "infinite gain" will distinguish this filter from the multiple-feedback filters that we consider later in Sec. 2.8.

Analysis of Fig. 2-7 shows that the infinite-gain MFB filter achieves Eq. (2.7) with

$$b_0 = \frac{1}{R_2 R_3 C C_1}$$

$$b_1 = \frac{1}{C}\left(\frac{1}{R_1} + \frac{1}{R_2} + \frac{1}{R_3}\right) \tag{2.9}$$

$$G = -\frac{R_2}{R_1}$$

We note that the filter has an *inverting* (negative) gain of R_2/R_1 and hence is useful if a phase shift of 180° is desired near $\omega = 0$.

The infinite-gain MFB filter is a very popular inverting-gain type. Some of its advantages include a minimal number of network elements (it requires one less resistor per stage than the VCVS of the previous section), good stability characteristics, and low output impedance [28].

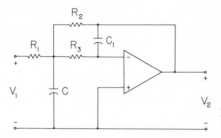

Figure 2-7. A second-order infinite-gain multiple-feedback low-pass filter.

For rapid design of infinite-gain MFB low-pass filters, the reader should select a value of C, calculate the K parameter by Eq. (2.6) or read it from one of Figs. 2-16a, b, or c, and use the appropriate one of Tables 2-26 through 2-49 at the end of the chapter to find the circuit element values. A summary of the design procedure including practical suggestions is given with the circuit in Sec. 2.11.

2.7 Biquad Low-Pass Filters

A circuit that requires more elements than the VCVS or the infinite-gain MFB filters of the previous two sections, but that has significant tuning advantages and excellent stability is the so-called *biquad* circuit [29] of Fig. 2-8. This circuit achieves Eq. (2.7), for both C and R_4 normalized to 1, with

$$b_0 = \frac{1}{R_3}$$

$$b_1 = \frac{1}{R_2} \tag{2.10}$$

$$G = \frac{R_3}{R_1}$$

These same results follow except that the gain is inverting if the output V_2 is taken at point a in Fig. 2-8.

Although the biquad filter requires a relatively large number of elements as compared to the VCVS and infinite-gain MFB filters, it is a very widely used design. The primary reason for this is its excellent tuning features and good stability, which are particularly important in the cascading of several stages [30].

The general design procedure for the biquad low-pass filter is given in Sec. 2.12. The design tables for determining element values are Tables 2-50 through 2-53.

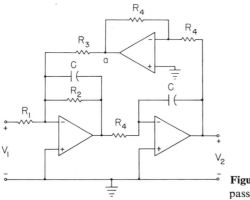

Figure 2-8.   A second-order biquad low-pass filter.

2.8   Multiple-Feedback Low-Pass Filters

Low-pass filters of our own design [31] are similar to cascaded VCVS filters with the added feature of feedback through capacitors from the output of each op-amp. These circuits, which we refer to as *multiple-feedback* low-pass filters, are collected in this section for orders $n = 3$ to 8. With the capacitor C of the circuits and the cutoff frequency ω_c both normalized to 1, the transfer function (2.2) is achieved with the equations given below. To obtain the normalized circuit element values, these equations were then iterated on a digital computer.

For $n = 3$, the circuit is shown in Fig. 2-9, and Eq. (2.2) is achieved with

$$b_0 = \frac{1}{R_1 R_2 R_3 C_1 C_2}$$

$$b_1 = A\left(\frac{1}{R_1} + \frac{1}{R_2}\right) + \frac{1}{R_2 R_3 C_2} - \frac{1}{R_2{}^2 C_1} \qquad (2.11)$$

$$b_2 = A + \frac{1}{R_1} + \frac{1}{R_2}$$

$$G = \mu$$

where

$$\mu = 1 + \frac{R_5}{R_4}$$

$$A = \left[\frac{C_1}{R_3}(1 - \mu) + C_2\left(\frac{1}{R_2} + \frac{1}{R_3}\right)\right]\Big/C_1 C_2 \qquad (2.12)$$

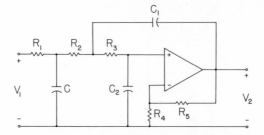

Figure 2-9.   A third-order multiple-feedback low-pass filter.

For $n = 4$, we use Fig. 2-10, for which

$$b_0 = \frac{1}{2R_1R_2R_3R_4}$$

$$b_1 = \frac{A}{2R_3R_4} + \frac{B}{2R_1R_2} - \frac{\mu}{2R_2R_3R_4}$$

$$b_2 = \frac{1}{R_3R_4} + \frac{1}{2R_1R_2} + \frac{AB}{2} \qquad\qquad (2.13)$$

$$b_2 = \frac{A}{2} + B$$

$$G = \mu$$

where

$$\mu = 1 + \frac{R_6}{R_5}$$

$$A = \frac{1}{R_1} + \frac{2}{R_2} \qquad\qquad (2.14)$$

$$B = \frac{1}{R_3} + \frac{1}{R_4}(2 - \mu)$$

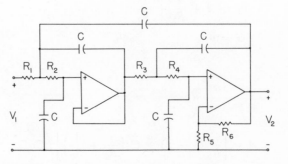

Figure 2-10.   A fourth-order multiple-feedback low-pass filter.

2.8 MULTIPLE-FEEDBACK LOW-PASS FILTERS

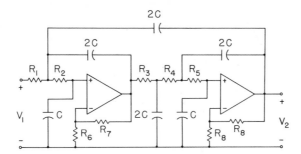

Figure 2-11. A fifth-order multiple-feedback low-pass filter.

For $n = 5$, we use Fig. 2-11, for which

$$b_0 = \frac{1}{16R_1R_2R_3R_4R_5}$$

$$b_1 = \frac{D}{16R_4R_5} + \frac{EF}{16} - \frac{\mu}{4R_2R_3R_4R_5} - \frac{1}{16R_4{}^2}\left(\frac{1}{R_1R_2} + \frac{A}{R_5}\right)$$

$$b_2 = \frac{B}{16R_4R_5} + \frac{DF}{16} + \frac{E}{8} - \frac{1}{16R_4{}^2}\left(A + \frac{4}{R_5}\right) \qquad (2.15)$$

$$b_3 = \frac{1}{2R_4R_5} + \frac{BF}{16} + \frac{D}{8} - \frac{1}{4R_4{}^2}$$

$$b_4 = \frac{F}{2} + \frac{B}{8}$$

$$G = 2\mu$$

where

$$\mu = 1 + \frac{R_7}{R_6}$$

$$A = \frac{1}{R_1} + \frac{1}{R_2}(5 - 2\mu)$$

$$B = 2A + 4\left(\frac{1}{R_3} + \frac{1}{R_4}\right)$$

$$D = \frac{2}{R_1R_2} + A\left(\frac{1}{R_3} + \frac{1}{R_4}\right) \qquad (2.16)$$

$$E = \frac{1}{R_1R_2}\left(\frac{1}{R_3} + \frac{1}{R_4}\right)$$

$$F = \frac{1}{R_4} - \frac{1}{R_5}$$

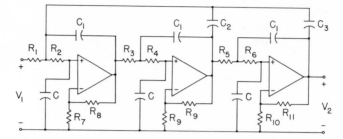

Figure 2-12. A sixth-order multiple-feedback low-pass filter.

For $n = 6$, we use Fig. 2-12, for which

$$b_0 = \frac{1}{R_1 R_2 R_3 R_4 R_5 R_6 A C_1{}^2}$$

$$b_1 = \frac{1}{AC_1{}^2}\left[\left(\frac{B}{R_3 R_4} + \frac{D}{R_1 R_2}\right)\frac{1}{R_5 R_6} + \frac{E}{R_1 R_2 R_3 R_4}\right.$$
$$\left. - \frac{2\mu_1}{R_2 R_3 R_4 R_5 R_6}(C_3 \mu_2 + C_2)\right]$$

$$b_2 = \frac{1}{AC_1{}^2}\left[\left(\frac{A}{R_3 R_4} + BD + \frac{C_1}{R_1 R_2}\right)\frac{1}{R_5 R_6} + \left(\frac{B}{R_3 R_4} + \frac{D}{R_1 R_2}\right)E\right.$$
$$\left. + \frac{1}{R_2 R_3 R_4}\left(\frac{C_1}{R_1} - 2C_2 \mu_1 E\right)\right] \qquad (2.17)$$

$$b_3 = \frac{1}{AC_1{}^2}\left[(AD + BC_1)\frac{1}{R_5 R_6} + \left(\frac{A}{R_3 R_4} + BD + \frac{C_1}{R_1 R_2}\right)E\right.$$
$$\left. + \left(\frac{B}{R_3 R_4} + \frac{D}{R_1 R_2}\right)C_1 - \frac{2C_1 C_2 \mu_1}{R_2 R_3 R_4}\right]$$

$$b_4 = \frac{1}{AC_1{}^2}\left[\frac{AC_1}{R_5 R_6} + (AD + BC_1)E + \left(\frac{A}{R_3 R_4} + BD + \frac{C_1}{R_1 R_2}\right)C_1\right]$$

$$b_5 = \frac{1}{AC_1{}^2}[AC_1 E + (AD + BC_1)C_1]$$

$$G = 2\mu_1 \mu_2$$

where

$$\mu_1 = 1 + \frac{R_8}{R_7}$$

$$\mu_2 = 1 + \frac{R_{11}}{R_{10}}$$

$$A = C_1 + C_2 + C_3 \tag{2.18}$$

$$B = \frac{1}{R_1} + \frac{1}{R_2}(A + 1 - C_1\mu_1)$$

$$D = \frac{1}{R_3} + \frac{1}{R_4}(1 - C_1)$$

$$E = \frac{1}{R_5} + \frac{1}{R_6}[1 + C_1(1 - \mu_2)]$$

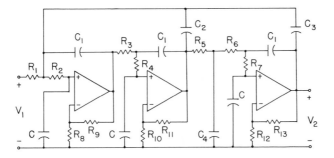

Figure 2-13. A seventh-order multiple-feedback low-pass filter.

For $n = 7$, we use Fig. 2-13, for which

$$b_0 = \frac{E}{BC_1{}^2C_4R_1R_2R_3R_4}$$

$$b_1 = \frac{1}{BC_1{}^2C_4}\left[E\left(\frac{L}{R_3R_4} + \frac{F}{R_1R_2}\right) + \frac{D}{R_1R_2R_3R_4}\right.$$
$$\left. - \frac{\mu_1\mu_2}{R_2R_3R_4}\left(\frac{\mu_3C_3}{R_5R_6R_7} + C_2E\right)\right]$$

$$b_2 = \frac{1}{BC_1{}^2C_4}\left[E\left(\frac{B}{R_3R_4} + LF + \frac{C_1}{R_1R_2}\right) + D\left(\frac{L}{R_3R_4} + \frac{F}{R_1R_2}\right)\right.$$
$$\left. + \frac{H}{R_1R_2R_3R_4} - \frac{C_2\mu_1\mu_2D}{R_2R_3R_4}\right] \tag{2.19}$$

$$b_3 = \frac{1}{BC_1{}^2C_4}\left[E(BF + LC_1) + D\left(\frac{B}{R_3R_4} + LF + \frac{C_1}{R_1R_2}\right)\right.$$
$$\left. + H\left(\frac{L}{R_3R_4} + \frac{F}{R_1R_2}\right) + \frac{C_1C_4}{R_1R_2R_3R_4} - \frac{C_2\mu_1\mu_2H}{R_2R_3R_4}\right]$$

$$b_4 = \frac{1}{BC_1{}^2C_4} \left[BC_1E + D(BF + LC_1) + H\left(\frac{B}{R_3R_4} + LF + \frac{C_1}{R_1R_2}\right) \right.$$

$$\left. + C_1C_4\left(\frac{L}{R_3R_4} + \frac{F}{R_1R_2}\right) - \frac{C_1C_2C_4\mu_1\mu_2}{R_2R_3R_4} \right]$$

$$b_5 = \frac{1}{BC_1{}^2C_4} \left[BC_1D + H(BF + LC_1) + C_1C_4\left(\frac{B}{R_3R_4} + LF + \frac{C_1}{R_1R_2}\right) \right]$$

$$b_6 = \frac{1}{BC_1{}^2C_4} [BC_1H + C_1C_4(BF + LC_1)]$$

$$G = \mu_1\mu_2\mu_3$$

where

$$\mu_1 = 1 + \frac{R_9}{R_8}$$

$$\mu_2 = 1 + \frac{R_{11}}{R_{10}}$$

$$\mu_3 = 1 + \frac{R_{13}}{R_{12}}$$

$$A = \frac{1}{R_6} + \frac{1}{R_7}[1 + C_1(1 - \mu_3)]$$

$$B = C_1 + C_2 + C_3$$

$$D = \frac{C_4}{R_6R_7} + A\left(\frac{1}{R_5} + \frac{1}{R_6}\right) - \frac{1}{R_6{}^2}$$

$$E = \frac{1}{R_6R_7}\left(\frac{1}{R_5} + \frac{1}{R_6}\right) - \frac{1}{R_6{}^2R_7} \qquad (2.20)$$

$$F = \frac{1}{R_3} + \frac{1}{R_4}[1 + C_1(1 - \mu_2)]$$

$$H = C_1\left(\frac{1}{R_5} + \frac{1}{R_6}\right) + AC_4$$

$$L = \frac{1}{R_1} + \frac{1}{R_2}(1 + B - C_1\mu_1)$$

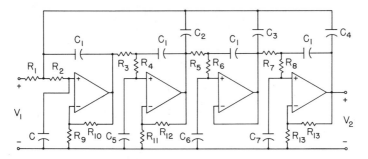

Figure 2-14.   An eighth-order multiple-feedback low-pass filter.

For $n = 8$, we use Fig. 2-14, for which

$$b_0 = \frac{1}{T R_1 R_2 R_3 R_4 R_5 R_6 R_7 R_8}$$

$$b_1 = \frac{1}{T} \left[\frac{M}{R_5 R_6 R_7 R_8} + \frac{Q}{R_1 R_2 R_3 R_4} - \frac{\mu_1 \mu_2}{R_2 R_3 R_4 R_5 R_6 R_7 R_8} (C_2 + C_3 + 2C_4) \right]$$

$$b_2 = \frac{1}{T} \left[\frac{L}{R_5 R_6 R_7 R_8} + MQ + \frac{P}{R_1 R_2 R_3 R_4} - \frac{\mu_1 \mu_2}{R_2 R_3 R_4} \left(C_2 Q + \frac{C_3 A}{R_5 R_6} \right) \right]$$

$$b_3 = \frac{1}{T} \left[\frac{H}{R_5 R_6 R_7 R_8} + LQ + MP + \frac{N}{R_1 R_2 R_3 R_4} \right.$$

$$\left. - \frac{\mu_1 \mu_2}{R_2 R_3 R_4} \left(C_2 P + \frac{C_1 C_3 C_7}{R_5 R_6} \right) \right] \qquad (2.21)$$

$$b_4 = \frac{1}{T} \left(\frac{E C_1 C_6}{R_5 R_6 R_7 R_8} + HQ + LP + MN + \frac{C_1{}^2 C_5 C_7}{R_1 R_2 R_3 R_4} - \frac{\mu_1 \mu_2 C_2 N}{R_2 R_3 R_4} \right)$$

$$b_5 = \frac{1}{T} \left(E C_1 C_6 Q + HP + LN + C_1{}^2 C_5 C_7 M - \frac{\mu_1 \mu_2 C_1{}^2 C_2 C_5 C_7}{R_2 R_3 R_4} \right)$$

$$b_6 = \frac{1}{T} \left(E C_1 C_6 P + HN + C_1{}^2 C_5 C_7 L \right)$$

$$b_7 = \frac{1}{T} \left(E C_1 C_6 N + C_1{}^2 C_5 C_7 H \right)$$

$$G = 2 \mu_1 \mu_2$$

where

$$\mu_1 = 1 + \frac{R_{10}}{R_9}$$

$$\mu_2 = 1 + \frac{R_{12}}{R_{11}}$$

$$A = C_7 \left(\frac{1}{R_7} + \frac{1}{R_8} \right) - \frac{C_1}{R_8}$$

$$B = C_5 \left(\frac{1}{R_5} + \frac{1}{R_6} \right)$$

$$D = C_6 \left(\frac{1}{R_3} + \frac{1}{R_4} \right) + \frac{C_1}{R_4} (1 - \mu_2)$$

$$E = C_1 + C_2 + C_3 + C_4$$

$$F = \frac{1}{R_2} (1 + E - C_1 \mu_1) + \frac{1}{R_1} \qquad (2.22)$$

$$H = ED + FC_1C_6$$

$$L = \frac{E}{R_3 R_4} + FD + \frac{C_1 C_6}{R_1 R_2}$$

$$M = \frac{F}{R_3 R_4} + \frac{D}{R_1 R_2}$$

$$N = AC_1 C_5 + BC_1 C_7$$

$$P = \frac{C_1 C_5}{R_7 R_8} + AB + \frac{C_1 C_7}{R_5 R_6}$$

$$Q = \frac{B}{R_7 R_8} + \frac{A}{R_5 R_6}$$

$$T = EC_1{}^3 C_5 C_6 C_7$$

An obvious difficulty of the multiple-feedback circuits of this section is the mathematical complexity of solving the various sets of equations for each filter type. For the cases we have given, solutions have been found using an optimization routine on a digital computer. The advantages are those of a multiple-feedback structure over a cascaded structure. Some of these are that the multiple-feedback structure's transfer function is less sensitive to changes in the subnetworks, the feedback from various stages gives somewhat better stability, and the circuits are somewhat easier to tune to the desired response, particularly with the inclusion of VCVS elements [31], [32]. We have spent considerable time in the laboratory constructing breadboard models of both multiple-feedback and cascaded higher-order filters, and it is our firm belief that our multiple-feedback circuits are far superior.

The general design procedure for the multiple-feedback low-pass filters is given in Sec. 2.13. The circuit element values may be obtained from Tables 2-54 through 2-61, which follow the design procedure. Examples of filters that were actually constructed are given in the next section.

2.9 Examples of Multiple-Feedback Low-Pass Filters

As an example of the procedure for obtaining multiple-feedback low-pass filters as described in the previous section, suppose it is desired to construct a sixth-order $\frac{1}{2}$ dB Chebyshev low-pass filter with $f_c = 2000$ Hz and $G = 2$. Selecting C in Fig. 2-12 to be 0.01 μF, we have by Eq. (2.6) (or by Fig. 2-16b) a K parameter of 5. The circuit element values are given in Table 2-59 where the resistances are given for a K parameter of 1 and therefore must be multiplied by 5. The results are $R_1 = 7.06$, $R_2 = 8.55$, $R_3 = 9.17$, $R_4 = 2.21$, $R_5 = 23.25$, $R_6 = 12.53$, $R_7 = R_{10} =$ open circuit, $R_8 = R_{11} = 0$; $R_9 = 22.70$ (all kilohms) and $C_1 = C_2 = 1.5C = 0.015 \, \mu$F, $C_3 = 0.33C = 0.0033 \, \mu$F. Using these values of capacitances with resistances as close as possible to the calculated values, the circuit was constructed with the amplitude response shown in Fig. 2-15d. The top half of the response was also shown earlier in Fig. 1-1b. The actual results were $G = 2$ (set by tuning the gain of the second VCVS), $f_c = 1972$ Hz, and a decibel ripple of 0.54.

To illustrate the improvement as the order increases, we have collected six low-pass responses for $n = 3, 4, 5, 6, 7$, and 8 in Fig. 2-15. These responses also illustrate the superiority of the cutoff features of the Chebyshev filter over those of the Butterworth filter. Other than the sixth-order Chebyshev response of Fig. 2-15d, the responses shown are (a) a third-order 3 dB Chebyshev with $f_c = 1000$ Hz and $G = 2$, (b) a fourth-order 1 dB Chebyshev with $f_c = 2000$ Hz and $G = 3$, (c) a fifth-order 0.1 dB Chebyshev with $f_c = 2000$ Hz and $G = 4$, (e) a seventh-order 0.5 dB Chebyshev with $f_c = 1000$ Hz and $G = 2$, and (f) an eighth-order Butterworth with $f_c = 1000$ Hz and $G = 4$.

2.10 Summary of VCVS Low-Pass Filter Design Procedure

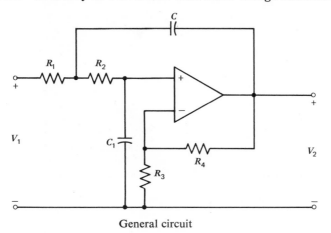

General circuit

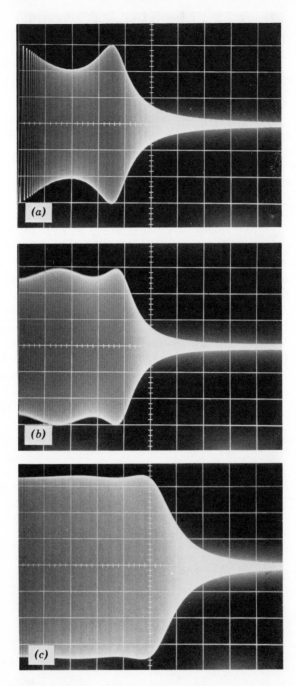

Figure 2-15. (a) A third-order, 3 dB Chebyshev; (b) a fourth-order, 1 dB Chebyshev; (c) a fifth-order, 0.1 dB Chebyshev; (d) a sixth-order, 0.5 dB Chebyshev; (e) a seventh-order, 0.5 dB Chebyshev; and (f) an eighth-order, Butterworth low-pass response.

24

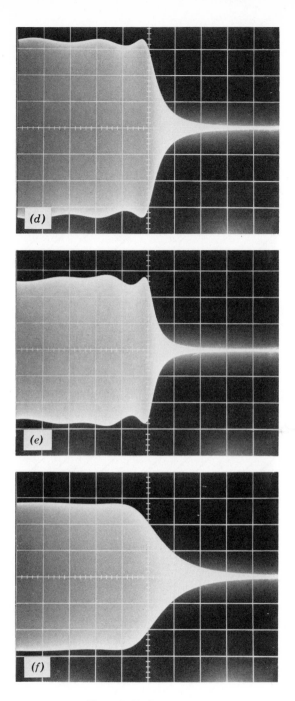

Figure 2-15. (*Continued*)

25

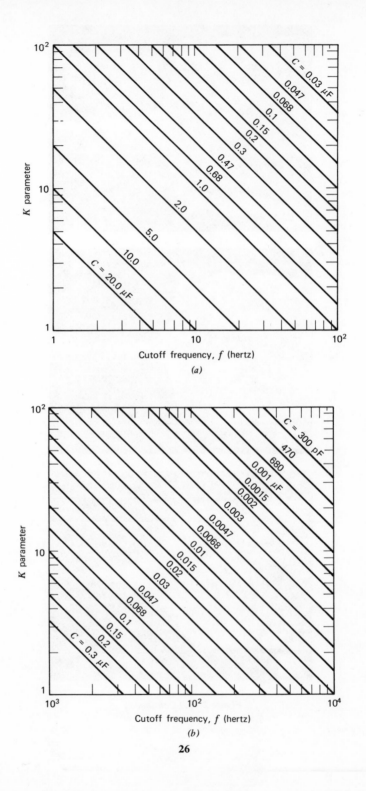

Cutoff frequency, f (hertz)

(a)

Cutoff frequency, f (hertz)

(b)

26

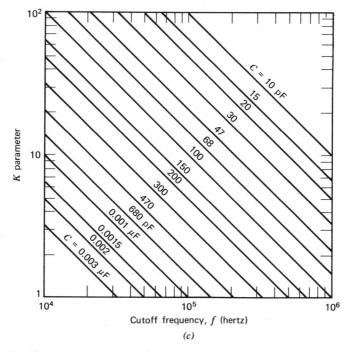

(c)

Figure 2-16. (a) K parameter versus frequency. (b) K parameter versus frequency. (c) K parameter versus frequency.

Procedure

Given cutoff f_c (hertz), gain G, order n, and filter type (Butterworth or Chebyshev), perform the following steps for a second-order filter, or for each stage of a higher-order cascaded filter ($n = 4, 6, 8$).

1.   Select a value of capacitance C and determine a K parameter from

$$K = \frac{100}{f_c C'}$$

where C' is the value of C in microfarads. Alternatively K may be found from Fig. 2-16a, b, or c. For higher-order designs (say, $n > 4$), it is better to use the equation since greater accuracy is required.

2.   Find the remaining element values from the appropriate one of Tables 2-2 through 2-25 as follows. The values of C_1 are determined directly from the tables using the chosen value of C. The resistances in the tables are given for $K = 1$, and hence their values must be multiplied by the K parameter of step 1 to yield the resistances of the circuit.

3. Select standard resistance values that are as close as possible to those indicated by the table and construct the filter, or its stages, in accordance with the general circuit. In case C_1 is a multiple of C such as 0.47, and so forth, standard values of C_1 result if C is chosen as a power of 10 (i.e., 0.1, 1, 10, etc.) μF.

Comments and Suggestions

(a) In the case of multiple-stage filters ($n > 2$), the K parameter for each stage need not be the same. That is, one may select a different C for each stage and using the given f_c of the overall filter calculate a K parameter for each stage. The resistances for each stage are those of the table times the K parameter of the stage.

(b) For best performance, the input resistance of the op-amp should be at least 10 times $R_{eq} = R_1 + R_2$. For a specific op-amp, this condition can generally be satisfied by proper selection of C to obtain a suitable K parameter.

(c) The values in the tables for R_3 and R_4 were determined to minimize the dc offset of the op-amp. Other values of R_3 and R_4 may be used as long as the ratio R_4/R_3 is the same as that of the table values.

(d) Standard resistance values of 5% tolerance normally yield acceptable results in the lower-order cases. For fifth and sixth orders, resistances of 2% tolerance probably should be used, and for seventh and eighth orders, 1% tolerance resistances probably should be used. In all cases, for best performance, resistance values close to those indicated by the tables should be used.

In the case of capacitors, percentage tolerances should parallel those given above for the resistors for best results. Since precision capacitors are relatively expensive, it may be desirable to use capacitors of higher tolerances, in which case trimming is generally required. In the case of low orders ($n \leq 4$), 10% capacitors are quite often satisfactory.

(e) The gain of each stage of the filter is $1 + R_4/R_3$, which can be adjusted to the correct value by using a potentiometer in lieu of resistors R_3 and R_4. This is accomplished by connecting the center tap of the potentiometer to the inverting input of the op-amp. These gain adjustments are very useful in tuning the overall response of the filter.

(f) Finally, there must be a dc return to ground at the filter input, the open-loop gain of the op-amp should be at least 50 times the gain of the filter at f_c, and the desired peak-to-peak voltage at f_c should not exceed $10^6/\pi f_c$ times the slew rate of the op-amp. Thus for high values of f_c, externally compensated op-amps may be required.

A specific example of a second-order VCVS design was given in Sec. 2.5.

Table 2-2. Second-Order Low-Pass Butterworth
VCVS Filter Designs

	Circuit Element Values[a]					
Gain	1	2	4	6	8	10
R_1	1.422	1.126	0.824	0.617	0.521	0.462
R_2	5.399	2.250	1.537	2.051	2.429	2.742
R_3	Open	6.752	3.148	3.203	3.372	3.560
R_4	0	6.752	9.444	16.012	23.602	32.038
C_1	0.33C	C	2C	2C	2C	2C

[a] Resistances in kilohms for a K parameter of 1.

Table 2-3. Second-Order Low-Pass Chebyshev
VCVS Filter Designs (0.1 dB)

	Circuit Element Values[a]					
Gain	1	2	4	6	8	10
R_1	0.912	0.671	0.474	0.349	0.293	0.259
R_2	2.541	1.139	0.807	1.094	1.303	1.476
R_3	Open	3.621	1.707	1.732	1.824	1.928
R_4	0	3.621	5.122	8.660	12.771	17.351
C_1	0.33C	C	2C	2C	2C	2C

[a] Resistances in kilohms for a K parameter of 1.

Table 2-4. Second-Order Low-Pass Chebyshev
VCVS Filter Designs (0.5 dB)

	Circuit Element Values[a]					
Gain	1	2	4	6	8	10
R_1	1.407	1.116	0.613	0.537	0.447	0.393
R_2	5.398	1.497	2.724	1.555	1.867	2.125
R_3	Open	5.226	4.450	2.510	2.645	2.798
R_4	0	5.226	13.349	12.552	18.517	25.179
C_1	0.22C	C	C	2C	2C	2C

[a] Resistances in kilohms for a K parameter of 1.

Table 2-5. Second-Order Low-Pass Chebyshev
VCVS Filter Designs (1 dB)

Gain	Circuit Element Values[a]					
	1	2	4	6	8	10
R_1	2.010	1.449	0.918	0.650	0.538	0.471
R_2	5.198	1.586	1.252	1.768	2.137	2.441
R_3	Open	6.071	2.893	2.902	3.057	3.235
R_4	0	6.071	8.680	14.508	21.400	29.117
C_1	0.22C	C	2C	2C	2C	2C

[a] Resistances in kilohms for a K parameter of 1.

Table 2-6. Second-Order Low-Pass Chebyshev
VCVS Filter Designs (2 dB)

Gain	Circuit Element Values[a]					
	1	2	4	6	8	10
R_1	2.328	1.980	1.141	0.786	0.644	0.561
R_2	13.220	1.555	1.348	1.957	2.388	2.742
R_3	Open	7.069	3.320	3.292	3.466	3.670
R_4	0	7.069	9.959	16.460	24.261	33.031
C_1	0.1C	C	2C	2C	2C	2C

[a] Resistances in kilohms for a K parameter of 1.

Table 2-7. Second-Order Low-Pass Chebyshev
VCVS Filter Designs (3 dB)

Gain	Circuit Element Values[a]					
	1	2	4	6	8	10
R_1	3.153	2.468	1.301	0.877	0.713	0.619
R_2	11.346	1.450	1.375	2.040	2.508	2.891
R_3	Open	7.835	3.568	3.500	3.682	3.900
R_4	0	7.835	10.704	17.502	25.771	35.096
C_1	0.1C	C	2C	2C	2C	2C

[a] Resistances in kilohms for a K parameter of 1.

Table 2-8. Fourth-Order Low-Pass Butterworth Cascaded VCVS Filter Designs

Gain	Circuit Element Values[a]						Stage
	1	2	6	10	36	100	
R_1	2.661	2.079	1.095	0.738	0.738	0.521	
R_2	9.521	1.218	2.313	3.432	1.716	2.432	
R_3	Open	6.595	5.112	5.213	2.945	3.281	1
R_4	0	6.595	10.225	20.851	14.725	29.527	
C_1	0.1C	C	C	C	2C	2C	
R_1	1.048	1.048	0.861	0.861	0.551	0.427	
R_2	4.833	4.833	2.941	2.941	2.297	2.965	
R_3	Open	Open	7.604	7.604	3.418	3.769	2
R_4	0	0	7.604	7.604	17.092	33.924	
C_1	0.5C	0.5C	C	C	2C	2C	

[a] Resistances in kilohms for a K parameter of 1.

Table 2-9. Fourth-Order Low-Pass Chebyshev Cascaded VCVS Filter Designs (0.1 dB)

Gain	Circuit Element Values[a]						Stage
	1	2	6	10	36	100	
R_1	4.555	3.013	1.100	0.698	0.698	0.478	
R_2	8.897	0.632	1.732	2.727	1.364	1.991	
R_3	Open	7.290	4.248	4.282	2.475	2.743	1
R_4	0	7.290	8.495	17.129	12.373	24.691	
C_1	0.047C	C	C	C	2C	2C	
R_1	1.632	1.632	1.248	1.248	0.742	0.564	
R_2	5.302	5.302	3.259	3.259	2.742	3.604	
R_3	Open	Open	9.014	9.014	4.180	4.632	2
R_4	0	0	9.014	9.014	20.901	41.684	
C_1	0.47C	0.47C	C	C	2C	2C	

[a] Resistances in kilohms for a K parameter of 1.

Table 2-10. Fourth-Order Low-Pass Chebyshev Cascaded VCVS Filter Design (0.5 dB)

Gain	Circuit Element Values[a]						Stage
	1	2	6	10	36	100	
R_1	7.220	4.538	1.303	0.808	0.808	0.547	
R_2	12.219	0.525	1.828	2.948	1.474	2.177	
R_3	Open	10.126	4.696	4.695	2.738	3.027	1
R_4	0	10.126	9.393	18.781	13.692	27.240	
C_1	0.027C	C	C	C	2C	2C	
R_1	2.994	2.994	1.880	1.880	1.033	0.773	
R_2	5.050	0.050	3.781	3.781	3.440	4.596	
R_3	Open	Open	11.321	11.321	5.368	5.966	2
R_4	0	0	11.321	11.321	26.838	53.695	
C_1	0.47C	0.47C	C	C	2C	2C	

[a] Resistances in kilohms for a K parameter of 1.

Table 2-11. Fourth-Order Low-Pass Chebyshev Cascaded VCVS Filter Designs (1 dB)

Gain	Circuit Element Values[a]						Stage
	1	2	6	10	36	100	
R_1	7.656	1.860	1.167	0.853	0.853	0.574	
R_2	22.360	2.761	4.401	3.010	1.505	2.235	
R_3	Open	9.242	8.352	4.829	2.830	3.122	1
R_4	0	9.242	16.708	19.314	14.148	28.097	
C_1	0.015C	0.5C	0.5C	C	2C	2C	
R_1	3.297	3.297	2.362	2.362	1.100	0.896	
R_2	8.333	8.333	3.838	3.838	8.239	5.057	
R_3	Open	Open	12.400	12.400	11.208	6.614	2
R_4	0	0	12.400	12.400	56.038	59.530	
C_1	0.33C	0.33C	C	C	C	2C	

[a] Resistances in kilohms for a K parameter of 1.

Table 2-12. Fourth-Order Low-Pass Chebyshev Cascaded VCVS Filter Designs (2 dB)

Gain	Circuit Element Values[a]						Stage
	1	2	6	10	36	100	
R_1	10.879	2.004	1.234	0.895	0.895	0.599	
R_2	25.072	2.723	4.421	3.046	1.523	2.277	
R_3	Open	9.453	8.482	4.927	2.902	3.195	1
R_4	0	9.453	16.964	19.707	14.511	28.756	
C_1	0.01C	0.5C	0.5C	C	2C	2C	
R_1	4.219	4.219	3.143	3.143	1.438	1.044	
R_2	12.316	12.316	3.638	3.638	3.976	5.474	
R_3	Open	Open	13.561	13.561	6.496	7.242	2
R_4	0	0	13.561	13.561	32.480	65.181	
C_1	0.22C	0.22C	C	C	2C	2C	

[a] Resistances in kilohms for a K parameter of 1.

Table 2-13. Fourth-Order Low-Pass Chebyshev Cascaded VCVS Filter Designs (3 dB)

Gain	Circuit Element Values[a]						Stage
	1	2	6	10	36	100	
R_1	11.575	2.087	1.271	0.918	0.918	0.612	
R_2	48.465	2.688	4.414	3.055	1.527	2.292	
R_3	Open	9.550	8.527	4.966	2.935	3.226	1
R_4	0	9.550	17.054	19.865	14.673	29.038	
C_1	0.005C	0.5C	0.5C	C	2C	2C	
R_1	6.094	6.094	3.870	3.870	1.592	1.141	
R_2	10.604	10.604	3.340	3.340	4.058	5.664	
R_3	Open	Open	14.420	14.420	6.781	7.561	2
R_4	0	0	14.420	14.420	33.905	68.046	
C_1	0.2C	0.2C	C	C	2C	2C	

[a] Resistances in kilohms for a K parameter of 1.

Table 2-14. Sixth-Order Low-Pass Butterworth Cascaded VCVS Filter Designs

Gain	Circuit Element Values[a]						Stage
	1	4	10	50	100	500	
R_1	4.090	3.075	1.573	0.792	0.792	0.546	
R_2	12.387	0.824	1.610	3.199	3.199	2.321	
R_3	Open	7.797	5.306	4.989	4.989	3.186	1
R_4	0	7.797	7.958	19.955	19.955	28.670	
C_1	0.05C	C	C	C	C	2C	
R_1	1.808	1.125	1.125	0.696	0.696	0.462	
R_2	2.981	2.251	2.251	1.821	1.821	2.742	
R_3	Open	6.752	6.752	3.146	3.146	3.560	2
R_4	0	6.752	6.752	12.582	12.582	32.039	
C_1	0.47C	C	C	2C	2C	2C	
R_1	0.980	0.980	0.824	0.824	0.675	0.594	
R_2	5.169	5.169	3.075	3.075	1.875	2.132	
R_3	Open	Open	7.797	7.797	3.401	3.407	3
R_4	0	0	7.797	7.797	10.202	13.628	
C_1	0.5C	0.5C	C	C	2C	2C	

[a] Resistances in kilohms for a K parameter of 1.

Table 2-15. Sixth-Order Low-Pass Chebyshev Cascaded VCVS Filter Designs (0.1 dB)

Gain	Circuit Element Values[a]						Stage
	1	4	10	50	100	500	
R_1	10.083	1.819	1.345	0.812	0.812	0.543	
R_2	22.240	2.466	3.336	2.760	2.760	2.063	
R_3	Open	8.569	7.800	4.466	4.466	2.897	1
R_4	0	8.569	11.701	17.864	17.864	26.069	
C_1	0.01C	0.5C	0.5C	C	C	2C	
R_1	3.300	2.539	2.539	1.037	1.037	0.626	
R_2	11.023	1.432	1.432	1.753	1.753	2.906	
R_3	Open	7.943	7.943	3.488	3.488	3.924	2
R_4	0	7.943	7.943	13.952	13.952	35.319	
C_1	0.1C	C	C	2C	2C	2C	
R_1	2.429	2.429	1.859	1.859	1.252	1.125	
R_2	7.917	7.917	5.173	5.173	7.678	8.548	
R_3	Open	Open	14.064	14.064	11.908	12.092	3
R_4	0	0	14.064	14.064	35.723	48.366	
C_1	0.5C	0.5C	C	C	C	C	

[a] Resistances in kilohms for a K parameter of 1.

35

Table 2-16. Sixth-Order Low-Pass Chebyshev Cascaded VCVS Filter Designs (0.5 dB)

Gain	Circuit Element Values[a]						Stage
	1	4	10	50	100	500	
R_1	14.752	1.997	1.457	1.054	1.054	0.578	
R_2	33.569	2.480	3.398	1.175	1.175	2.143	
R_3	Open	8.954	8.092	2.786	2.786	3.023	1
R_4	0	8.954	12.138	11.143	11.143	27.206	
C_1	0.005C	0.5C	0.5C	2C	2C	2C	
R_1	4.727	4.727	3.751	1.021	1.021	0.706	
R_2	18.163	1.145	1.145	4.206	4.206	3.042	
R_3	Open	9.791	9.791	6.534	6.534	4.164	2
R_4	0	9.791	9.791	26.135	26.135	37.477	
C_1	0.05C	C	C	C	C	2C	
R_1	4.374	4.374	2.746	2.746	1.729	1.538	
R_2	7.377	7.377	5.876	5.876	9.333	10.490	
R_3	Open	Open	17.243	17.243	14.749	15.035	3
R_4	0	0	17.243	17.243	44.247	60.139	
C_1	0.5C	0.5C	C	C	C	C	

[a] Resistances in kilohms for a K parameter of 1.

Table 2-17. Sixth-Order Low-Pass Chebyshev Cascaded VCVS Filter Designs (1 dB)

Gain	Circuit Element Values[a]						Stage
	1	4	10	50	100	500	
R_1	18.375	2.070	1.502	0.890	0.890	0.590	
R_2	42.164	2.470	3.404	2.871	2.871	2.166	
R_3	Open	9.080	8.177	4.702	4.702	3.062	1
R_4	0	9.080	12.265	18.808	18.808	27.560	
C_1	0.0033C	0.5C	0.5C	C	C	2C	
R_1	7.191	4.684	4.684	1.284	1.284	0.739	
R_2	13.438	0.970	0.970	1.769	1.769	3.072	
R_3	Open	11.308	11.308	3.816	3.816	4.235	2
R_4	0	11.308	11.308	15.263	15.263	38.112	
C_1	0.047C	C	C	2C	2C	2C	
R_1	4.616	4.616	3.429	3.429	2.033	1.796	
R_2	13.333	13.333	5.923	5.923	9.990	11.311	
R_3	Open	Open	18.705	18.705	16.031	16.383	3
R_4	0	0	18.705	18.705	48.092	65.533	
C_1	0.33C	0.33C	C	C	C	C	

[a] Resistances in kilohms for a K parameter of 1.

Table 2-18. Sixth-Order Low-Pass Chebyshev Cascaded VCVS Filter Designs (2 dB)

Gain	Circuit Element Values[a]						Stage
	1	4	10	50	100	500	
R_1	25.047	2.141	1.544	1.107	1.107	0.601	
R_2	52.349	2.450	3.397	1.184	1.184	2.181	
R_3	Open	9.181	8.235	2.864	2.864	3.092	1
R_4	0	9.181	12.353	11.457	11.457	27.824	
C_1	0.002C	0.5C	0.5C	2C	2C	2C	
R_1	9.150	2.411	2.411	1.137	1.137	0.771	
R_2	19.239	3.944	3.944	4.179	4.179	3.082	
R_3	Open	12.708	12.708	6.645	6.645	4.281	2
R_4	0	12.708	12.708	26.581	26.581	38.531	
C_1	0.027C	0.5C	0.5C	C	C	2C	
R_1	6.728	6.728	4.539	4.539	2.965	2.428	
R_2	13.955	13.955	5.584	5.584	4.275	5.220	3
R_3	Open	Open	20.247	20.247	9.653	9.560	
R_4	0	0	20.247	20.247	28.959	38.241	
C_1	0.27C	0.27C	C	2C	2C	2C	

[a] Resistances in kilohms for a K parameter of 1.

38

Table 2-19. Sixth-Order Low-Pass Chebyshev Cascaded VCVS Filter Designs (3 dB)

Gain	Circuit Element Values[a]						Stage
	1	4	10	50	100	500	
R_1	28.422	2.179	1.566	0.919	0.919	0.607	
R_2	77.782	2.434	3.387	2.886	2.886	2.187	
R_3	Open	9.228	8.256	4.756	4.756	3.104	1
R_4	0	9.228	12.384	19.025	19.025	27.933	
C_1	0.0012C	0.5C	0.5C	C	C	2C	
R_1	12.613	2.543	2.543	1.170	1.170	0.788	
R_2	19.243	3.817	3.817	4.148	4.148	3.078	
R_3	Open	12.721	12.721	6.648	6.648	4.296	2
R_4	0	12.721	12.721	26.591	26.591	38.667	
C_1	0.02C	0.5C	0.5C	C	C	2C	
R_1	8.220	8.220	5.578	5.578	3.364	2.709	
R_2	17.350	17.350	5.114	5.114	4.239	5.266	
R_3	Open	Open	21.383	21.383	10.138	9.968	3
R_4	0	0	21.383	21.383	30.414	39.870	
C_1	0.2C	0.2C	C	C	2C	2C	

[a] Resistances in kilohms for a K parameter of 1.

Table 2-20. Eighth-Order Low-Pass Butterworth Cascaded VCVS Filter Designs

Gain	Circuit Element Values[a]						Stage
	1	4	10	50	100	500	
R_1	5.978	1.714	1.311	0.981	0.981	0.981	
R_2	12.840	2.956	3.864	1.291	1.291	1.291	
R_3	Open	9.340	8.625	2.840	2.840	2.840	1
R_4	0	9.340	12.938	11.361	11.361	11.361	
C_1	0.033C	0.5C	0.5C	2C	2C	2C	
R_1	1.865	1.432	1.432	0.767	0.767	0.767	
R_2	6.173	1.768	1.768	1.651	1.651	1.651	
R_3	Open	6.402	6.402	3.023	3.023	3.023	2
R_4	0	6.402	6.402	12.091	12.091	12.091	
C_1	0.22C	C	C	2C	2C	2C	
R_1	1.254	1.254	0.957	0.957	0.957	0.644	
R_2	4.039	4.039	2.647	2.647	2.647	1.967	
R_3	Open	Open	7.207	7.207	7.207	3.264	3
R_4	0	0	7.207	7.207	7.207	13.055	
C_1	0.5C	0.5C	C	C	C	2C	
R_1	0.959	0.959	0.959	0.959	0.811	0.668	
R_2	5.285	5.285	5.285	5.285	3.122	1.895	
R_3	Open	Open	Open	Open	7.867	3.418	4
R_4	0	0	0	0	7.867	10.254	
C_1	0.5C	0.5C	0.5C	0.5C	C	2C	

[a] Resistances in kilohms for a K parameter of 1.

Table 2-21. Eighth-Order Low-Pass Chebyshev Cascaded VCVS Filter Designs (0.1 dB)

Gain	Circuit Element Values[a]						Stage
	1	4	10	50	100	500	
R_1	18.140	1.994	1.994	0.857	0.857	0.857	
R_2	39.560	2.375	1.188	2.762	2.762	2.762	
R_3	Open	8.738	5.303	4.524	4.524	4.524	1
R_4	0	8.738	7.954	18.098	18.098	18.098	
C_1	0.0033C	0.5C	C	C	C	C	
R_1	6.009	1.895	1.895	1.091	1.091	1.091	
R_2	15.989	3.346	3.346	1.454	1.454	1.454	
R_3	Open	10.482	10.482	3.180	3.180	3.180	2
R_4	0	10.482	10.482	12.720	12.720	12.720	
C_1	0.033C	0.5C	0.5C	2C	2C	2C	
R_1	4.167	4.167	2.918	2.918	2.918	1.299	
R_2	9.734	9.734	2.085	2.085	2.085	2.342	
R_3	Open	Open	10.007	10.007	10.007	4.551	3
R_4	0	0	10.007	10.007	10.007	18.205	
C_1	0.15C	0.15C	C	C	C	2C	
R_1	3.204	3.204	3.204	3.204	2.474	1.939	
R_2	10.858	10.858	10.858	10.858	7.031	4.485	
R_3	Open	Open	Open	Open	19.009	8.565	4
R_4	0	0	0	0	19.009	25.696	
C_1	0.5C	0.5C	0.5C	0.5C	C	2C	

[a] Resistances in kilohms for a K parameter of 1.

41

Table 2-22. Eighth-Order Low-Pass Chebyshev Cascaded VCVS Filter Designs (0.5 dB)

Gain	Circuit Element Values[a]						Stage
	1	4	10	50	100	500	
R_1	25.168	2.104	2.104	1.085	1.085	1.085	
R_2	66.305	2.379	1.189	1.154	1.154	1.154	
R_3	Open	8.967	5.490	2.798	2.798	2.798	1
R_4	0	8.967	8.235	11.193	11.193	11.193	
C_1	0.0015C	0.5C	C	2C	2C	2C	
R_1	8.382	2.135	2.135	1.181	1.181	1.181	
R_2	27.176	3.201	3.201	1.447	1.447	1.447	
R_3	Open	10.672	10.672	3.285	3.285	3.285	2
R_4	0	10.672	10.672	13.139	13.139	13.139	
C_1	0.015C	0.5C	0.5C	2C	2C	2C	
R_1	5.550	5.550	4.280	4.280	4.280	1.284	
R_2	18.714	18.714	1.650	1.650	1.650	5.501	
R_3	Open	Open	11.861	11.861	11.861	8.481	3
R_4	0	0	11.861	11.861	11.861	33.926	
C_1	0.068C	0.068C	C	C	C	C	
R_1	5.623	5.623	5.623	5.623	3.629	2.705	
R_2	10.232	10.232	10.232	10.232	7.927	5.316	
R_3	Open	Open	Open	Open	23.113	10.696	4
R_4	0	0	0	0	23.113	32.088	
C_1	0.5C	0.5C	0.5C	0.5C	C	2C	

[a] Resistances in kilohms for a K parameter of 1.

42

Table 2.23. Eighth-Order Low-Pass Chebyshev Cascaded VCVS Filter Designs (1 dB)

Gain	Circuit Element Values[a]						Stage
	1	4	10	50	100	500	
R_1	31.691	2.148	2.148	1.101	1.101	1.101	
R_2	80.400	2.372	1.186	1.157	1.157	1.157	
R_3	Open	9.041	5.557	2.823	2.823	2.823	1
R_4	0	9.041	8.336	11.291	11.291	11.291	
C_1	0.001C	0.5C	C	2C	2C	2C	
R_1	11.777	2.244	2.244	1.218	1.218	1.218	
R_2	24.772	3.121	3.121	1.437	1.437	1.437	
R_3	Open	10.729	10.729	3.319	3.319	3.319	2
R_4	0	10.729	10.729	13.276	13.276	13.276	
C_1	0.012C	0.5C	0.5C	2C	2C	2C	
R_1	7.187	7.187	5.333	5.333	5.333	1.610	
R_2	20.680	20.680	1.393	1.393	1.393	2.307	
R_3	Open	Open	13.454	13.454	13.454	4.897	3
R_4	0	0	13.454	13.454	13.454	19.588	
C_1	0.05C	0.05C	C	C	C	2C	
R_1	6.023	6.023	6.023	6.023	4.521	3.221	
R_2	18.139	18.139	18.139	18.139	7.973	5.597	
R_3	Open	Open	Open	Open	24.990	11.757	4
R_4	0	0	0	0	24.990	35.270	
C_1	0.33C	0.33C	0.33C	0.33C	C	2C	

[a] Resistances in kilohms for a K parameter of 1.

Table 2-24. Eighth-Order Low-Pass Chebyshev Cascaded VCVS Filter Designs (2 dB)

Gain	Circuit Element Values[a]						Stage
	1	4	10	50	100	500	
R_1	49.085	2.189	2.189	1.115	1.115	1.115	
R_2	77.409	2.361	1.180	1.158	1.158	1.158	
R_3	Open	9.099	5.615	2.842	2.842	2.842	1
R_4	0	9.099	8.423	11.368	11.368	11.368	
C_1	0.00068C	0.5C	C	2C	2C	2C	
R_1	15.178	2.355	2.355	1.254	1.254	1.254	
R_2	34.578	3.031	3.031	1.423	1.423	1.423	
R_3	Open	10.772	10.772	3.346	3.346	3.346	2
R_4	0	10.772	10.772	13.385	13.385	13.385	
C_1	0.0068C	0.5C	0.5C	2C	2C	2C	
R_1	10.130	10.130	7.048	7.048	7.048	1.712	
R_2	23.166	23.166	1.099	1.099	1.099	2.262	
R_3	Open	Open	16.293	16.293	16.293	4.967	3
R_4	0	0	16.293	16.293	16.293	19.868	
C_1	0.033C	0.033C	C	C	C	2C	
R_1	9.869	9.869	9.869	9.869	5.975	3.923	
R_2	15.142	15.142	15.142	15.142	7.503	5.713	
R_3	Open	Open	Open	Open	26.957	12.849	4
R_4	0	0	0	0	26.957	38.547	
C_1	0.3C	0.3C	0.3C	0.3C	C	2C	

[a] Resistances in kilohms for a K parameter of 1.

Table 2-25. Eighth-Order Low-Pass Chebyshev Cascaded VCVS Filter Designs (3 dB)

Gain	Circuit Element Values[a]						Stage
	1	4	10	50	100	500	
R_1	65.326	2.211	2.211	1.123	1.123	1.123	
R_2	84.688	2.352	1.176	1.158	1.158	1.158	
R_3	Open	9.126	5.645	2.851	2.851	2.851	1
R_4	0	9.126	8.468	11.404	11.404	11.404	
C_1	0.00047C	0.5C	C	2C	2C	2C	
R_1	20.532	2.420	2.420	1.274	1.274	1.274	
R_2	35.070	2.976	2.976	1.413	1.413	1.413	
R_3	Open	10.791	10.791	3.359	3.359	3.359	2
R_4	0	10.791	10.791	13.435	13.435	13.435	
C_1	0.005C	0.5C	0.5C	2C	2C	2C	
R_1	12.304	12.304	3.164	3.164	3.164	1.477	
R_2	29.162	29.162	4.989	4.989	4.989	5.344	
R_3	Open	Open	16.307	16.307	16.307	8.526	3
R_4	0	0	16.307	16.307	16.307	34.105	
C_1	0.022C	0.022C	0.5C	0.5C	0.5C	C	
R_1	11.789	11.789	11.789	11.789	7.336	4.451	
R_2	19.419	19.419	19.419	19.419	6.866	5.658	
R_3	Open	Open	Open	Open	28.403	13.478	4
R_4	0	0	0	0	28.403	40.435	
C_1	0.22C	0.22C	0.22C	0.22C	C	2C	

[a] Resistances in kilohms for a K parameter of 1.

2.11   Summary of Infinite-Gain MFB Low-Pass Filter Design Procedure

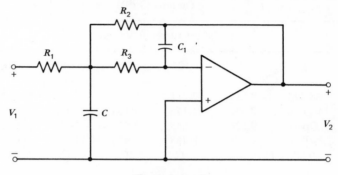

General circuit

Procedure

Given cutoff f_c (hertz), gain G, order n, and filter type (Butterworth or Chebyshev), perform the following steps for a second-order filter, or for each stage of a higher-order cascaded filter ($n = 4, 6, 8$).

1.   Select a value of capacitance C and determine a K parameter from

$$K = \frac{100}{f_c C'}$$

where C' is the value of C in microfarads. Alternatively, K may be found from Fig. 2-16a, b, or c. For higher-order designs (say $n > 4$), it is better to use the equation since greater accuracy is required.

2.   Find the remaining element values from the appropriate one of Tables 2-26 through 2-49 as follows. The values of C_1 are determined directly from the tables using the chosen value of C. The resistances in the tables are given for $K = 1$ and hence their values must be multiplied by the K parameter of step 1 to yield the resistances of the circuit.

3.   Select standard resistance values that are as close as possible to those indicated by the table and construct the filter, or its stages, in accordance with the general circuit. In case C_1 is a multiple of C such as 0.47, and so forth, standard values of C_1 result if C is chosen as a power of 10 (i.e., 0.1, 1, 10, etc.) μF.

Comments and Suggestions

The comments and suggestions for the VCVS low-pass filter given in Sec. 2.10 apply as follows:

(a) Paragraphs (a) and (d) are directly applicable.
(b) Paragraphs (c) and (e) do not apply.

(c) Paragraph (b) applies with the exception that

$$R_{eq} = R_3 + \frac{R_1 R_2}{R_1 + R_2}$$

(d) Paragraph (f) applies except that the dc return to ground is already satisfied by R_2 and R_3.

In addition, the following applies:

(e) The inverting gain of each stage of the filter is R_2/R_1. Gain adjustments can be made by using a potentiometer in lieu of R_2.

(f) For minimum dc offset, a resistance equal to R_{eq} of (c) can be placed in the noninverting input to ground.

The infinite-gain MFB low-pass filter was discussed in Sec. 2.6.

Table 2-26. Second-Order Low-Pass Butterworth Infinite-Gain MFB Filter Designs

	Circuit Element Values[a]			
Gain	1	2	6	10
R_1	3.111	2.565	1.697	1.625
R_2	3.111	5.130	10.180	16.252
R_3	4.072	3.292	4.977	4.723
C_1	0.2C	0.15C	0.05C	0.033C

[a] Resistances in kilohms for a K parameter of 1.

Table 2-27. Second-Order Low-Pass Chebyshev Infinite-Gain MFB Filter Designs (0.1 dB)

	Circuit Element Values[a]			
Gain	1	2	6	10
R_1	2.163	1.306	1.103	1.069
R_2	2.163	2.611	6.619	10.690
R_3	1.767	2.928	2.310	2.167
C_1	0.2C	0.1C	0.05C	0.033C

[a] Resistances in kilohms for a K parameter of 1.

Table 2-28. Second-Order Low-Pass Chebyshev Infinite-Gain MFB Filter Designs (0.5 dB)

Gain	Circuit Element Values[a]			
	1	2	6	10
R_1	3.374	2.530	1.673	1.608
R_2	3.374	5.060	10.036	16.083
R_3	3.301	3.301	5.045	4.722
C_1	0.15C	0.1C	0.033C	0.022C

[a] Resistances in kilohms for a K parameter of 1.

Table 2-29. Second-Order Low-Pass Chebyshev Infinite-Gain MFB Filter Designs (1 dB)

Gain	Circuit Element Values[a]			
	1	2	6	10
R_1	3.821	2.602	2.284	2.213
R_2	3.821	5.204	13.705	22.128
R_3	6.013	8.830	5.588	5.191
C_1	0.1C	0.05C	0.03C	0.02C

[a] Resistances in kilohms for a K parameter of 1.

Table 2-30. Second-Order Low-Pass Chebyshev Infinite-Gain MFB Filter Designs (2 dB)

Gain	Circuit Element Values[a]			
	1	2	6	10
R_1	4.658	3.999	3.009	3.113
R_2	4.658	7.997	18.053	31.133
R_3	13.216	7.697	8.524	6.591
C_1	0.05C	0.05C	0.02C	0.015C

[a] Resistances in kilohms for a K parameter of 1.

Table 2-31. Second-Order Low-Pass Chebyshev Infinite-Gain MFB Filter Designs (3 dB)

	Circuit Element Values[a]			
Gain	1	2	6	10
R_1	6.308	6.170	3.754	3.617
R_2	6.308	12.341	22.524	36.171
R_3	11.344	6.169	10.590	9.892
C_1	0.05C	0.047C	0.015C	0.01C

[a] Resistances in kilohms for a K parameter of 1.

Table 2-32. Fourth-Order Low-Pass Butterworth Cascaded MFB Filter Designs

	Circuit Element Values[a]				
Gain	1	4	36	100	Stage
R_1	5.321	5.230	3.167	3.052	
R_2	5.321	10.460	19.003	30.522	1
R_3	9.521	5.153	8.886	8.299	
C_1	0.05C	0.047C	0.015C	0.01C	
R_1	2.334	1.750	1.411	1.187	
R_2	2.334	3.501	8.467	11.871	2
R_3	3.289	3.289	2.992	4.268	
C_1	0.33C	0.22C	0.1C	0.05C	

[a] Resistances in kilohms for a K parameter of 1.

Table 2-33. Fourth-Order Low-Pass Chebyshev Cascaded MFB Filter Designs (0.1 dB)

Gain	Circuit Element Values[a]				Stage
	1	4	36	100	
R_1	8.102	6.563	4.458	4.262	
R_2	8.102	13.126	26.747	42.623	
R_3	11.755	9.675	14.243	13.542	1
C_1	0.02C	0.015C	0.005C	0.0033C	
R_1	3.886	2.914	1.732	1.965	
R_2	3.886	5.829	10.392	19.653	2
R_3	3.489	3.489	7.827	4.139	
C_1	0.3C	0.2C	0.05C	0.05C	

[a] Resistances in kilohms for a K parameter of 1.

Table 2-34. Fourth-Order Low-Pass Chebyshev Cascaded MFB Filter Designs (0.5 dB)

Gain	Circuit Element Values[a]				Stage
	1	4	36	100	
R_1	11.672	9.821	7.311	7.112	
R_2	11.672	19.643	43.866	71.125	
R_3	20.405	14.787	16.453	15.221	1
C_1	0.01C	0.0082C	0.0033C	0.0022C	
R_1	5.178	3.449	2.827	2.990	
R_2	5.178	6.898	16.964	29.898	
R_3	6.863	10.304	8.379	6.095	2
C_1	0.2C	0.1C	0.05C	0.039C	

[a] Resistances in kilohms for a K parameter of 1.

Table 2-35. Fourth-Order Low-Pass Chebyshev Cascaded MFB Filter Designs (1 dB)

Gain	Circuit Element Values[a]				Stage
	1	4	36	100	
R_1	16.165	11.483	9.060	8.930	
R_2	16.165	22.967	54.361	89.302	1
R_3	19.371	22.360	21.470	19.168	
C_1	0.0082C	0.005C	0.0022C	0.0015C	
R_1	6.252	4.689	4.018	3.919	
R_2	6.252	9.379	24.107	39.193	2
R_3	9.667	9.667	7.521	7.010	
C_1	0.15C	0.1C	0.05C	0.033C	

[a] Resistances in kilohms for a K parameter of 1.

Table 2-36. Fourth-Order Low-Pass Chebyshev Cascaded MFB Filter Designs (2 dB)

Gain	Circuit Element Values[a]				Stage
	1	4	36	100	
R_1	21.758	16.196	13.240	13.171	
R_2	21.758	32.392	79.438	131.710	1
R_3	25.072	25.517	22.891	20.709	
C_1	0.005C	0.0033C	0.0015C	0.001C	
R_1	8.080	6.796	5.060	4.922	
R_2	8.080	13.592	30.359	49.219	2
R_3	14.149	10.257	11.411	10.558	
C_1	0.1C	0.082C	0.033C	0.022C	

[a] Resistances in kilohms for a K parameter of 1.

Table 2-37. Fourth-Order Low-Pass Chebyshev Cascaded MFB
Filter Designs (3 dB)

Gain	Circuit Element Values[a]				Stage
	1	4	36	100	
R_1	26.277	19.708	16.048	16.285	
R_2	26.277	39.415	96.286	162.846	
R_3	32.346	32.346	29.130	25.329	1
C_1	0.0033C	0.0022C	0.001C	0.00068C	
R_1	12.189	7.481	6.679	5.734	
R_2	12.189	14.962	40.075	57.338	
R_3	10.604	17.277	11.945	15.028	2
C_1	0.1C	0.05C	0.027C	0.015C	

[a] Resistances in kilohms for a K parameter of 1.

Table 2-38. Sixth-Order Low-Pass Butterworth Cascaded
MFB Filter Designs

Gain	Circuit Element Values[a]				Stage
	1	4	36	100	
R_1	9.296	6.972	5.205	4.753	
R_2	9.296	13.944	31.230	47.534	
R_3	9.083	9.083	9.891	10.658	1
C_1	0.03C	0.02C	0.0082C	0.005C	
R_1	3.343	2.565	2.154	2.154	
R_2	3.343	5.130	12.924	12.924	
R_3	3.444	3.292	2.882	2.882	2
C_1	0.22C	0.15C	0.068C	0.068C	
R_1	2.139	2.139	2.139	2.129	
R_2	2.139	2.139	2.139	3.548	
R_3	3.589	3.589	3.589	2.163	3
C_1	0.33C	0.33C	0.33C	0.33C	

[a] Resistances in kilohms for a K parameter of 1.

Table 2-39. Sixth-Order Low-Pass Chebyshev Cascaded MFB Filter Designs (0.1 dB)

Gain	Circuit Element Values[a]				Stage
	1	4	36	100	
R_1	20.167	15.002	12.321	12.351	
R_2	20.167	30.005	73.926	123.511	1
R_3	22.240	22.648	20.223	18.157	
C_1	0.005C	0.0033C	0.0015C	0.001C	
R_1	8.548	4.928	3.936	3.936	
R_2	8.548	9.857	23.618	23.618	2
R_3	6.257	11.181	10.266	10.266	
C_1	0.068C	0.033C	0.015C	0.015C	
R_1	5.424	5.424	5.424	4.262	
R_2	5.424	5.424	5.424	7.103	3
R_3	5.910	5.910	5.910	6.154	
C_1	0.3C	0.3C	0.3C	0.22C	

[a] Resistances in kilohms for a K parameter of 1.

Table 2-40. Sixth-Order Low-Pass Chebyshev Cascaded MFB Filter Designs (0.5 dB)

Gain	Circuit Element Values[a]				Stage
	1	4	36	100	
R_1	31.791	20.685	16.622	17.915	
R_2	31.791	41.371	99.732	179.150	1
R_3	28.846	39.900	36.510	27.642	
C_1	0.0027C	0.0015C	0.00068C	0.0005C	
R_1	10.976	8.232	6.801	6.801	
R_2	10.976	16.463	40.808	40.808	2
R_3	11.853	11.853	10.521	10.521	
C_1	0.033C	0.022C	0.01C	0.01C	
R_1	7.728	7.728	7.728	5.849	
R_2	7.728	7.728	7.728	9.749	3
R_3	9.489	9.489	9.489	11.033	
C_1	0.22C	0.22C	0.22C	0.15C	

[a] Resistances in kilohms for a K parameter of 1.

Table 2-41. Sixth-Order Low-Pass Chebyshev Cascaded MFB Filter Designs (1 dB)

Gain	Circuit Element Values[a]				Stage
	1	4	36	100	
R_1	34.567	25.925	22.602	22.269	
R_2	34.567	51.850	135.610	222.690	1
R_3	49.310	49.310	37.707	34.791	
C_1	0.0015C	0.001C	0.0005C	0.00033C	
R_1	13.510	10.325	8.519	8.519	
R_2	13.510	20.650	51.112	51.112	2
R_3	15.281	14.663	13.067	13.067	
C_1	0.022C	0.015C	0.0068C	0.0068C	
R_1	10.787	10.787	10.787	8.630	
R_2	10.787	10.787	10.787	14.383	3
R_3	9.415	9.415	9.415	9.415	
C_1	0.2C	0.2C	0.2C	0.15C	

[a] Resistances in kilohms for a K parameter of 1.

Table 2-42. Sixth-Order Low-Pass Chebyshev Cascaded MFB Filter Designs (2 dB)

Gain	Circuit Element Values[a]				Stage
	1	4	36	100	
R_1	50.093	38.299	30.786	24.408	
R_2	50.093	76.598	184.713	244.085	1
R_3	52.349	50.345	47.322	71.623	
C_1	0.001C	0.00068C	0.0003C	0.00015C	
R_1	21.182	12.815	10.786	10.786	
R_2	21.182	25.630	64.715	64.715	2
R_3	14.959	22.615	18.832	18.832	
C_1	0.015C	0.0082C	0.0039C	0.0039C	
R_1	15.698	15.698	15.698	10.645	
R_2	15.698	15.698	15.698	17.741	3
R_3	10.765	10.765	10.765	14.288	
C_1	0.15C	0.15C	0.15C	0.1C	

[a] Resistances in kilohms for a K parameter of 1.

Table 2-43. Sixth-Order Low-Pass Chebyshev Cascaded MFB Filter Designs (3 dB)

Gain	Circuit Element Values[a]				Stage
	1	4	36	100	
R_1	62.421	54.708	37.589	33.403	
R_2	62.421	109.416	225.533	334.033	1
R_3	62.499	48.491	58.813	66.183	
C_1	0.00068C	0.0005C	0.0002C	0.00012C	
R_1	25.227	14.925	11.289	11.289	
R_2	25.227	29.850	67.737	67.737	2
R_3	19.243	32.525	35.832	35.832	
C_1	0.01C	0.005C	0.002C	0.002C	
R_1	16.440	16.440	16.440	14.694	
R_2	16.440	16.440	16.440	24.491	3
R_3	17.350	17.350	17.350	14.203	
C_1	0.1C	0.1C	0.1C	0.082C	

[a] Resistances in kilohms for a K parameter of 1.

Table 2-44. Eighth-Order Low-Pass Butterworth Cascaded MFB Filter Designs

Gain	Circuit Element Values[a]				Stage
	1	4	36	100	
R_1	11.174	8.380	7.415	7.415	
R_2	11.174	16.761	44.490	44.490	1
R_3	15.113	15.113	11.387	11.387	
C_1	0.015C	0.01C	0.005C	0.005C	
R_1	4.907	3.681	2.494	2.494	
R_2	4.907	7.361	14.966	14.966	2
R_3	3.441	3.441	4.340	4.340	
C_1	0.15C	0.1C	0.039C	0.039C	
R_1	2.808	2.808	2.808	1.827	
R_2	2.808	2.808	2.808	5.076	3
R_3	3.007	3.007	3.007	3.327	
C_1	0.3C	0.3C	0.3C	0.15C	
R_1	2.820	2.820	2.820	2.820	
R_2	2.820	2.820	2.820	2.820	4
R_3	1.911	1.911	1.911	1.911	
C_1	0.47C	0.47C	0.47C	0.47C	

[a] Resistances in kilohms for a K parameter of 1.

Table 2-45. Eighth-Order Low-Pass Chebyshev Cascaded MFB Filter Designs (0.1 dB)

Gain	Circuit Element Values[a]				Stage
	1	4	36	100	
R_1	33.958	25.469	22.452	22.452	
R_2	33.958	50.938	134.713	134.713	1
R_3	46.490	46.490	35.158	35.158	
C_1	0.0015C	0.001C	0.0005C	0.0005C	
R_1	11.438	9.594	7.294	7.294	
R_2	11.438	19.189	43.763	43.763	2
R_3	18.478	13.768	14.488	14.488	
C_1	0.015C	0.012C	0.005C	0.005C	
R_1	9.074	9.074	9.074	7.362	
R_2	9.074	9.074	9.074	20.449	3
R_3	8.178	8.178	8.178	6.331	
C_1	0.082C	0.082C	0.082C	0.047C	
R_1	7.097	7.097	7.097	7.097	
R_2	7.097	7.097	7.097	7.097	4
R_3	8.170	8.170	8.170	8.170	
C_1	0.3C	0.3C	0.3C	0.3C	

[a] Resistances in kilohms for a K parameter of 1.

Table 2-46. Eighth-Order Low-Pass Chebyshev Cascaded MFB Filter Designs (0.5 dB)

Gain	Circuit Element Values[a]				Stage
	1	4	36	100	
R_1	53.757	37.752	28.277	28.277	
R_2	53.757	75.504	169.660	169.660	1
R_3	56.786	66.305	73.770	73.770	
C_1	0.00082C	0.0005C	0.0002C	0.0002C	
R_1	21.393	12.573	11.469	11.469	
R_2	21.393	25.146	68.816	68.816	2
R_3	15.972	27.176	18.390	18.390	
C_1	0.01C	0.005C	0.0027C	0.0027C	
R_1	14.803	14.803	14.803	7.949	
R_2	14.803	14.803	14.803	22.081	3
R_3	10.152	10.152	10.152	15.993	
C_1	0.047C	0.047C	0.047C	0.02C	
R_1	9.567	9.567	9.567	9.567	
R_2	9.567	9.567	9.567	9.567	4
R_3	15.035	15.035	15.035	15.035	
C_1	0.2C	0.2C	0.2C	0.2C	

[a] Resistances in kilohms for a K parameter of 1.

Table 2-47. Eighth-Order Low-Pass Chebyshev Cascaded MFB Filter Designs
(1 dB)

Gain	Circuit Element Values[a]				Stage
	1	4	36	100	
R_1	63.382	55.636	38.294	38.294	
R_2	63.382	111.273	229.761	229.761	1
R_3	80.400	58.714	73.931	73.931	
C_1	0.0005C	0.00039C	0.00015C	0.00015C	
R_1	20.984	17.289	12.539	12.539	
R_2	20.984	34.578	75.231	75.231	2
R_3	33.367	25.960	31.023	31.023	
C_1	0.005C	0.0039C	0.0015C	0.0015C	
R_1	16.600	16.600	16.600	10.637	
R_2	16.600	16.600	16.600	29.548	3
R_3	14.922	14.922	14.922	16.767	
C_1	0.03C	0.03C	0.03C	0.015C	
R_1	14.473	14.473	14.473	14.473	
R_2	14.473	14.473	14.473	14.473	4
R_3	12.052	12.052	12.052	12.052	
C_1	0.2C	0.2C	0.2C	0.2C	

[a] Resistances in kilohms for a K parameter of 1.

Table 2-48. Eighth-Order Low-Pass Chebyshev Cascaded MFB Filter Designs (2 dB)

Gain	Circuit Element Values[a]				Stage
	1	4	36	100	
R_1	93.911	64.259	61.000	61.000	
R_2	93.911	128.518	365.997	365.997	
R_3	83.371	100.520	70.594	70.594	1
C_1	0.00033C	0.0002C	0.0001C	0.0001C	
R_1	28.094	21.071	18.144	18.144	
R_2	28.094	42.142	108.863	108.863	
R_3	42.342	42.342	32.782	32.782	2
C_1	0.003C	0.002C	0.001C	0.001C	
R_1	19.050	19.050	19.050	16.312	
R_2	19.050	19.050	19.050	45.310	
R_3	27.100	27.100	27.100	17.091	3
C_1	0.015C	0.015C	0.015C	0.01C	
R_1	20.934	20.934	20.934	20.934	
R_2	20.934	20.934	20.934	20.934	
R_3	21.416	21.416	21.416	21.416	4
C_1	0.1C	0.1C	0.1C	0.1C	

[a] Resistances in kilohms for a K parameter of 1.

Table 2-49.   Eighth-Order Low-Pass Chebyshev Cascaded MFB Filter Designs
(3 dB)

Gain	Circuit Element Values[a]				Stage
	1	4	36	100	
R_1	105.077	89.088	56.708	56.708	
R_2	105.077	178.175	340.249	340.249	1
R_3	123.728	97.289	152.840	152.840	
C_1	0.0002C	0.00015C	0.00005C	0.00005C	
R_1	36.365	27.712	19.007	19.007	
R_2	36.365	55.425	114.041	114.041	2
R_3	45.001	43.305	63.139	63.139	
C_1	0.0022C	0.0015C	0.0005C	0.0005C	
R_1	23.212	23.212	23.212	15.360	
R_2	23.212	23.212	23.212	42.667	3
R_3	34.007	34.007	34.007	37.002	
C_1	0.01C	0.01C	0.01C	0.005C	
R_1	21.246	21.246	21.246	21.246	
R_2	21.246	21.246	21.246	21.246	4
R_3	23.706	23.706	23.706	23.706	
C_1	0.1C	0.1C	0.1C	0.1C	

[a] Resistances in kilohms for a K parameter of 1.

2.12   Summary of Biquad Low-Pass Filter Design Procedure

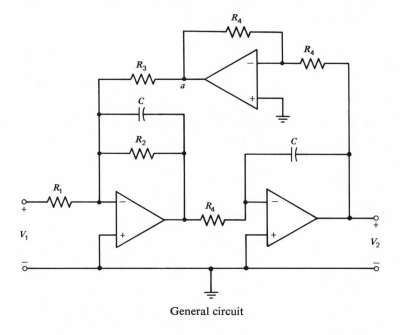

General circuit

Procedure

Given cutoff f_c (hertz), filter gain, order n, and filter type (Butterworth or Chebyshev), perform the following steps for a second-order filter, or for each stage of a higher-order cascaded filter ($n = 4, 6, 8$).

1.   Select a value of capacitance C and determine a K parameter from

$$K = \frac{100}{f_c C'}$$

where C' is the value of C in microfarads. Alternatively, K may be found from Fig. 2-16a, b, or c. For higher-order designs (say, $n > 4$), it is better to use the equation since greater accuracy is required.

2.   Find the resistance values from the appropriate one of Tables 2-50 through 2-53. The resistances in the tables are given for $K = 1$ and hence their values must be multiplied by the K parameter of step 1 to yield the resistances of the circuit. The number G in the tables is the stage gain, and for $n > 2$, the filter gain is the product of the stage gains. The stage gains, which are not necessarily equal, are chosen by the designer.

3. Select standard resistance values that are as close as possible to those indicated by the table and construct the filter, or its stages, in accordance with the general circuit.

Comments and Suggestions

The comments and suggestions for the VCVS low-pass filter given in Sec. 2.10 apply as follows:

(a) Paragraphs (a) and (d) are directly applicable.
(b) Paragraphs (c) and (e) do not apply.
(c) Paragraph (b) applies with the exception that $R_{eq} = R_1$ or R_4 since R_1 or R_4 is the resistor connected in the inverting input of the op-amps.
(d) Paragraph (f) applies except that the dc return to ground is already satisfied by R_2 and R_3.

In addition, the following applies:

(e) The stage gain is R_3/R_1. If an inverting gain of the same magnitude is desired, the output can be taken at point a.
(f) The filter response is readily tuned by varying R_1, R_2, and R_3. Varying R_1 affects the gain, varying R_2 affects the passband response, and varying R_3 changes f_c.

The biquad low-pass filters were discussed in Sec. 2.7.

Table 2-50. Second-Order Low-Pass Biquad Filter Designs

		Circuit Element Values[a]				
			Chebyshev			
	Butterworth	0.1 dB	0.5 dB	1 dB	2 dB	3 dB
R_1	1.592/G	0.480/G	1.050/G	1.444/G	1.934/G	2.248/G
R_2	1.125	0.671	1.116	1.450	1.980	2.468
R_3	1.592	0.480	1.050	1.444	1.934	2.248
R_4	1.592	1.592	1.592	1.592	1.592	1.592

[a] Resistances in kilohms for a K parameter of 1, G = gain.

Table 2-51. Fourth-Order Low-Pass Cascaded Biquad Filter Designs

Circuit Element Values[a]

	Butterworth	Chebyshev 0.1 dB	0.5 dB	1 dB	2 dB	3 dB	Stage
R_1	1.592/G	1.197/G	1.496/G	1.613/G	1.714/G	1.762/G	
R_2	2.079	3.013	4.538	5.703	7.587	9.343	1
R_3	1.592	1.197	1.496	1.613	1.714	1.762	
R_4	1.592	1.592	1.592	1.592	1.592	1.592	
R_1	1.592/G	2.555/G	4.465/G	5.696/G	7.183/G	8.121/G	
R_2	0.861	1.248	1.880	2.362	3.143	3.870	2
R_3	1.592	2.555	4.465	5.696	7.183	8.121	
R_4	1.592	1.592	1.592	1.592	1.592	1.592	

[a] Resistances in kilohms for a K parameter of 1, G = stage gain.

Table 2-52. Sixth-Order Low-Pass Cascaded Biquad Filter Designs

Circuit Element Values[a]

	Butterworth	Chebyshev 0.1 dB	0.5 dB	1 dB	2 dB	3 dB	Stage
R_1	1.592/G	1.409/G	1.566/G	1.606/G	1.648/G	1.667/G	
R_2	3.075	6.938	10.248	12.798	16.941	20.816	1
R_3	1.592	1.409	1.556	1.606	1.648	1.667	
R_4	1.592	1.592	1.592	1.592	1.592	1.592	
R_1	1.592/G	2.285/G	2.697/G	2.854/G	2.986/G	3.050/G	
R_2	1.125	2.539	3.751	4.684	6.201	7.619	2
R_3	1.592	2.285	2.697	2.854	2.986	3.050	
R_4	1.592	1.592	1.592	1.592	1.592	1.592	
R_1	1.592/G	6.042/G	10.137/G	12.762/G	15.927/G	17.922/G	
R_2	0.824	1.859	2.746	3.429	4.539	5.578	3
R_3	1.592	6.042	10.137	12.762	15.927	17.922	
R_4	1.592	1.592	1.592	1.592	1.592	1.592	

[a] Resistances in kilohms for a K parameter of 1, G = stage gain.

Table 2-53. Eighth-Order Low-Pass Cascaded Biquad Filter Designs

			Circuit Element Values[a]				
				Chebyshev			
	Butterworth	0.1 dB	0.5 dB	1 dB	2 dB	3 dB	Stage
R_1	1.592/G	1.488/G	1.573/G	1.601/G	1.623/G	1.634/G	
R_2	4.079	12.437	18.243	22.731	30.038	36.879	1
R_3	1.592	1.488	1.573	1.601	1.623	1.634	
R_4	1.592	1.592	1.592	1.592	1.592	1.592	
R_1	1.592/G	1.992/G	2.147/G	2.200/G	2.242/G	2.262/G	
R_2	1.432	4.367	6.406	7.982	10.548	12.950	2
R_3	1.592	1.992	2.147	2.200	2.242	2.262	
R_4	1.592	1.592	1.592	1.592	1.592	1.592	
R_1	1.592/G	3.823/G	4.438/G	4.669/G	4.866/G	4.960/G	
R_2	0.957	2.918	4.280	5.333	7.048	8.653	3
R_3	1.592	3.823	4.438	4.669	4.866	4.960	
R_4	1.592	1.592	1.592	1.592	1.592	1.592	
R_1	1.592/G	10.929/G	18.075/G	22.652/G	28.169/G	31.645/G	
R_2	0.811	2.474	3.629	4.521	5.975	7.336	4
R_3	1.592	10.929	18.075	22.652	28.169	31.645	
R_4	1.592	1.592	1.592	1.592	1.592	1.592	

[a] Resistances in kilohms for a K parameter of 1, G = stage gain.

2.13 Summary of Multiple-Feedback Low-Pass Filter Design Procedure

The general circuits of the multiple-feedback low-pass filters are given for $n = 3$ through 8 in Figs. 2-9 and 2-14, respectively.

Procedure

Given cutoff f_c (hertz), gain G, order n, and filter type (Butterworth or Chebyshev), perform the following steps.

1. Select a value of capacitance C and determine a K parameter from

$$K = \frac{100}{f_c C'}$$

where C' is the value of C in microfarads. Alternatively, K may be found from Fig. 2-16a, b, or c. For higher-order designs (say, $n > 4$), it is better to use the equation since greater accuracy is required.

2. Find the remaining element values from the appropriate one of Tables 2-54 through 2-61 as follows. The values of the capacitances other than C are determined directly from the tables using the chosen value of C. The resistances in the tables are given for $K = 1$, and hence their values must be multiplied by the K parameter of step 1 to yield the resistances of the circuit.

3. Select standard resistance values that are as close as possible to those indicated by the table and construct the filter in accordance with the general circuit. In case the remaining capacitances are multiples of C such as 0.47 and so forth, standard values result if C is chosen as a power of 10 (i.e., 0.1, 1, 10, etc.) μF.

Comments and Suggestions

The comments and suggestions for the VCVS low-pass filter given in Sec. 2.10 apply as follows:

(a) Paragraphs (d) and (f) apply directly.

(b) Paragraph (a) does not apply.

(c) Paragraph (b) applies with the exception that R_{eq} for each op-amp is the sum of the two or three resistors connected to the noninverting input.

(d) Paragraph (c) applies for each op-amp except that R_3 and R_4 are replaced by the two resistors connected to the inverting input.

(e) Paragraph (e) applies for each VCVS that has gain-setting resistors. In each case R_3 and R_4 are replaced by these gain-setting resistors.

Specific examples of various-order multiple-feedback low-pass filters were given in Sec. 2.9.

Table 2-54. Third-Order Low-Pass Multiple-Feedback Filter Designs

Circuit Element Values[a]

| | Butterworth | | Chebyshev | | | | | | | | | |
| | | | 0.1 dB | | 0.5 dB | | 1 dB | | 2 dB | | 3 dB | |
Gain	1	2	1	2	1	2	1	2	1	2	1	2
R_1	1.639	2.491	1.661	2.122	2.556	2.987	3.345	3.621	4.351	4.647	5.338	5.612
R_2	11.697	2.339	14.515	2.678	23.994	4.421	10.624	5.800	25.033	8.051	48.310	10.149
R_3	2.103	0.692	1.702	0.433	2.041	0.427	3.977	0.391	2.580	0.330	2.079	0.283
R_4	Open	11.043	Open	10.465	Open	15.668	Open	19.623	Open	26.056	Open	32.087
R_5	0	11.043	0	10.465	0	15.668	0	19.623	0	26.056	0	32.087
C_1	C	C	1.2C	C	1.5C	C	1.5C	C	2.2C	C	3C	C
C_2	0.1C	C	0.05C	C	0.03C	C	0.039C	C	0.02C	C	0.01C	C

[a] Resistances in kilohms for a K parameter of 1.

Table 2-55. Fourth-Order Low-Pass Multiple-Feedback Filter Designs:
Butterworth and Chebyshev (0.1 dB)

| | Circuit Element Values[a] | | | | | |
| | Butterworth | | | Chebyshev (0.1 dB) | | |
Gain	2	6	10	2	6	10
R_1	0.531	0.431	0.390	0.727	0.540	0.477
R_2	3.439	5.476	7.026	3.641	6.003	7.751
R_3	2.441	0.535	0.367	5.850	0.540	0.365
R_4	0.719	2.544	3.190	0.250	2.210	2.872
R_5	6.321	3.695	3.952	12.199	3.300	3.597
R_6	6.321	18.474	35.567	12.199	16.501	32.371

[a] Resistances in kilohms for a K parameter of 1.

Table 2-56. Fourth-Order Low-Pass Multiple-Feedback Filter Designs:
Chebyshev (0.5 dB) and Chebyshev (1 dB)

| | Circuit Element Values[a] | | | | | |
| | Chebyshev (0.5 dB) | | | Chebyshev (1 dB) | | |
Gain	3	6	10	3	6	10
R_1	0.916	0.735	0.635	1.099	0.855	0.729
R_2	5.197	7.345	9.582	5.463	7.902	10.401
R_3	1.321	0.639	0.429	1.431	0.681	0.456
R_4	1.346	2.455	3.240	1.355	2.530	3.366
R_5	4.000	3.712	4.076	4.179	3.853	4.247
R_6	8.000	18.561	36.686	8.357	19.267	38.224

[a] Resistances in kilohms for a K parameter of 1.

Table 2-57.   Fourth-Order Low-Pass Multiple-Feedback Filter Designs: Chebyshev (2 dB) and Chebyshev (3 dB)

	Circuit Element Values^a					
	Chebyshev (2 dB)			Chebyshev (3 dB)		
Gain	3	6	10	3	6	10
R_1	1.357	1.008	0.843	1.559	1.116	0.920
R_2	5.558	8.337	11.129	5.490	8.486	11.456
R_3	1.543	0.721	0.481	1.608	0.742	0.493
R_4	1.340	2.575	3.458	1.317	2.582	3.488
R_5	4.324	3.955	4.376	4.387	3.988	4.423
R_6	8.649	19.773	39.382	8.775	19.941	39.811

^{a} Resistances in kilohms for a K parameter of 1.

Table 2-58. Fifth-Order Low-Pass Multiple-Feedback Filter Designs

				Circuit Element Values[a]					

Circuit Element Values[a]

	Butterworth		Chebyshev						
			0.1 dB		0.5 dB		1 dB	2 dB	3 dB
Gain	4	10	4	10	4	10	4	4	6
R_1	0.169	0.107	0.300	0.162	0.443	0.208	0.562	0.856	0.447
R_2	1.631	2.190	1.597	2.629	1.469	2.932	1.277	0.868	1.997
R_3	4.980	5.992	6.558	7.520	8.929	9.606	10.858	14.851	14.915
R_4	0.836	0.869	0.982	1.007	1.143	1.158	1.222	1.316	1.322
R_5	0.557	0.521	0.505	0.484	0.538	0.525	0.546	0.538	0.579
R_6	3.600	2.870	3.800	3.490	3.800	3.930	3.680	3.460	3.670
R_7	3.600	11.490	3.800	13.950	3.800	15.700	3.680	3.460	7.340
R_8	12.800	14.760	16.200	18.000	21.200	22.600	25.260	33.400	33.600

[a] Resistances in kilohms for a K parameter of 1.

Table 2-59. Sixth-Order Low-Pass Multiple-Feedback Filter Designs

Circuit Element Values[a]

	Butterworth		Chebyshev						
			0.1 dB		0.5 dB		1 dB	2 dB	3 dB
Gain	2	4	2	4	2	4	4	4	4
R_1	0.251	0.358	0.156	0.150	1.411	1.151	1.139	0.948	0.635
R_2	4.194	4.914	5.474	6.017	1.709	2.277	2.363	2.569	2.821
R_3	0.894	1.076	0.144	0.131	1.834	1.835	1.968	2.091	2.128
R_4	1.302	1.295	1.599	2.130	0.442	0.692	0.611	0.499	0.423
R_5	0.859	5.718	0.425	0.151	4.650	2.443	3.177	4.177	4.676
R_6	0.973	0.465	0.770	1.711	2.505	4.782	4.685	4.607	4.658
R_7	Open	Open	Open	Open	Open	7.000	7.000	7.040	6.860
R_8	0	0	0	0	0	7.000	7.000	7.040	6.860
R_9	4.440	4.740	3.500	4.460	4.540	5.100	5.160	5.200	5.100
R_{10}	Open	12.340	Open	3.700	Open	Open	Open	Open	Open
R_{11}	0	12.340	0	3.700	0	0	0	0	0
C_1	2C	C	10C	10C	1.5C	1.5C	1.5C	1.5C	1.5C
C_2	C	C	C	C	1.5C	0.22C	0.33C	0.68C	1.5C
C_3	C	0.5C	C	C	0.33C	0.22C	0.33C	0.68C	1.5C

[a] Resistances in kilohms for a K parameter of 1.

Table 2-60. Seventh-Order Low-Pass Multiple-Feedback Filter Designs

	Circuit Element Values[a]							
			Chebyshev					
	Butterworth		0.1 dB		0.5 dB		1 dB	2 dB	3 dB
Gain	2	4	2	4	2	4	4	4	4
R_1	1.091	0.360	0.449	1.304	0.539	0.727	0.236	1.317	0.520
R_2	1.018	2.326	0.582	1.048	1.594	0.512	2.580	3.545	4.142
R_3	0.695	0.813	1.136	2.685	0.435	0.550	0.354	4.993	4.158
R_4	0.767	0.831	0.895	1.111	0.671	2.991	0.812	0.454	0.343
R_5	2.096	1.940	0.996	3.232	0.867	0.941	1.729	1.061	0.980
R_6	1.920	2.932	2.028	0.907	1.997	0.885	1.305	2.943	3.886
R_7	0.610	0.501	0.664	0.742	0.809	1.375	0.984	0.682	0.972
R_8	Open	5.372	2.062	4.700	Open	2.500	5.600	Open	Open
R_9	0	5.372	2.062	4.700	0	2.500	5.600	0	0
R_{10}	2.925	3.288	Open	Open	2.212	Open	2.330	10.900	9.100
R_{11}	2.925	3.288	0	0	2.212	0	2.330	10.900	9.100
R_{12}	Open	Open	Open	9.762	Open	6.440	Open	9.360	11.740
R_{13}	0	0	0	9.762	0	6.440	0	9.360	11.740
C_1	2.2C	2C	4.7C	2C	3.9C	3.3C	5C	1.2C	1.2C
C_2	C	C	C	0.2C	4.7C	0.33C	3.3C	1.8C	5C
C_3	0.47C	C	C	0.2C	2.2C	1.5C	4.7C	0.33C	0.68C
C_4	C	C	4.7C	3C	10C	15C	6.8C	12C	15C

[a] Resistances in kilohms for a K parameter of 1.

Table 2-61.   Eighth-Order Low-Pass Multiple-Feedback Filter
Designs

	Circuit Element Values[a]		
		Chebyshev	
	Butterworth	0.1 dB	0.5 dB
Gain	4	4	4
R_1	0.772	1.020	1.377
R_2	5.164	1.259	0.110
R_3	4.581	0.577	2.013
R_4	2.099	2.253	0.270
R_5	2.083	0.943	2.208
R_6	1.874	2.398	0.673
R_7	1.933	1.262	0.628
R_8	2.474	0.205	0.138
R_9	Open	Open	2.974
R_{10}	0	0	2.974
R_{11}	13.360	5.660	Open
R_{12}	13.360	5.660	0
R_{13}	8.814	2.934	1.532
C_1	0.68C	3C	5.6C
C_2	0.22C	2C	3.3C
C_3	0.3C	0.82C	0.22C
C_4	0.25C	0.56C	0.56C
C_5	0.5C	0.47C	4.7C
C_6	0.5C	3.9C	4.7C
C_7	0.5C	2.7C	4.7C

[a] Resistances in kilohms for a K parameter of 1.

3

High-Pass Filters

3.1 General Theory

A high-pass filter passes high frequencies and attenuates low frequencies. An ideal high-pass amplitude response, represented by the broken line, is shown in Fig. 3-1 with a realizable approximation to the ideal, represented by the solid line. The stopband is the interval $0 < \omega < \omega_c$ and the passband is the interval $\omega > \omega_c$, where the cutoff point $\omega_c = 2\pi f_c$ is the value at which the amplitude attains $1/\sqrt{2}$ times its maximum value, as in the low-pass case. Also, as in the low-pass case, the amplitude may be plotted in decibels, in which case the amplitude at ω_c is 3 dB below its maximum value.

High-pass filter transfer functions may be obtained from low-pass filter transfer functions by replacing s by $1/s$ [21]. Therefore, an nth-order high-pass transfer function derived from that of the all-pole low-pass filter of Eq. (2.2) is given by

$$\frac{V_2}{V_1} = \frac{Gb_0}{S^n + b_{n-1}S^{n-1} + \cdots + b_1 S + b_0}\bigg|_{S=1/s} \tag{3.1}$$

Making the indicated substitution we have, after simplification,

$$\frac{V_2}{V_1} = \frac{Gs^n}{s^n + a_{n-1}s^{n-1} + \cdots + a_1 s + a_0} \tag{3.2}$$

where, noting that $b_n = 1$, we have

$$a_{n-i} = \frac{b_i}{b_0}; \quad i = 1, 2, \ldots, n \tag{3.3}$$

73

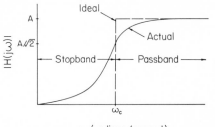

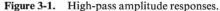

ω (radians/second)   **Figure 3-1.**   High-pass amplitude responses.

The *gain* of the high-pass filter is the value of its transfer function as s becomes infinite. In the case of Eq. (3.2), the gain is evidently G, which is also the gain of its low-pass "prototype" given by Eq. (3.1) before the substitution $S = 1/s$ is made.

The high-pass filter defined by Eq. (3.2) will be a Butterworth or a Chebyshev (with given decibel passband ripple) when its low-pass prototype is Butterworth or Chebyshev. In either case, normalized ($\omega_c = 1$) coefficients a_i given by Eq. (3.3) are obtained from the normalized coefficients b_i of the low-pass prototype. As in the low-pass case, ω_c is the conventional cutoff point in the Butterworth high-pass filter, and is the beginning of the ripple channel of the passband in the Chebyshev high-pass filter. (Of course, in the 3 dB case, ω_c is also the conventional cutoff point.) To relate ω_c (or f_c) for the Chebyshev case to the conventional cutoff point, the reader may consult Table 3-1 for various values of n and ripple widths.

The remarks in Secs. 2.2 and 2.3 of Chapter 2 concerning the advantages and disadvantages of Chebyshev and Butterworth low-pass filters also apply to high-pass filters. The Chebyshev has better cutoff properties with passband ripples, whereas the Butterworth has a better phase response and a monotonic

Table 3-1.   Ratio of Conventional Cutoff $f_{3\,dB}$ to Channel Terminal f_c
for High-Pass Chebyshev Filters

dB \ n	$f_{3\,dB}/f_c$						
	2	3	4	5	6	7	8
0.1	0.515	0.720	0.824	0.881	0.915	0.936	0.951
0.5	0.720	0.857	0.915	0.944	0.961	0.971	0.978
1	0.821	0.913	0.950	0.967	0.977	0.983	0.987
2	0.931	0.968	0.982	0.988	0.992	0.994	0.995
3	1.000	1.000	1.000	1.000	1.000	1.000	1.000

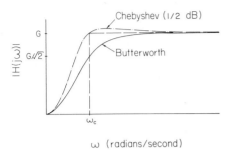
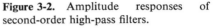

Figure 3-2. Amplitude responses of second-order high-pass filters.

amplitude response. Of course, the amplitude responses of each improve as the order n increases, with an attendant deterioration in the phase responses. Amplitude and phase responses of a Chebyshev and Butterworth filter of order 2 are shown for comparative purposes in Figs. 3-2 and 3-3.

As in the low-pass case, we have presented design tables for the rapid construction of high-pass filters of both Butterworth and Chebyshev type. The tables are located at the end of the chapter, and are designed to give resistance values for a K parameter of 1. The use of the tables for various high-pass filter circuits will be illustrated in the following sections of this chapter. The circuits considered are well-known and widely used high-pass filters, as well as one new type of our own design.

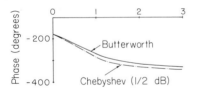

Figure 3-3. Phase responses of second-order high-pass filters.

3.2 VCVS High-Pass Filters

In the case $n = 2$, Eq. (3.2) becomes

$$\frac{V_2}{V_1} = \frac{Gs^2}{s^2 + a_1 s + a_0} \tag{3.4}$$

which is the transfer function of a second-order high-pass filter, or of a second-order stage of a higher-order high-pass filter. A widely used circuit which realizes this function is the *VCVS high-pass* filter of Fig. 3-4, credited to Sallen and Key [26]. It is obtained from the VCVS low-pass circuit of Fig. 2-6 by interchanging the position of the Cs and Rs (excluding R_3 and R_4),

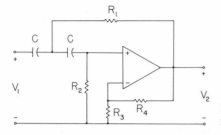

Figure 3-4. A second-order VCVS high-pass filter.

and contains a voltage-controlled-voltage-source (VCVS) consisting of the op-amp, R_3, and R_4.

Analysis of Fig. 3-4 shows that it achieves Eq. (3.4) with

$$a_0 = \frac{1}{R_1 R_2 C^2}$$

$$a_1 = \frac{1}{R_1 C}(1 - \mu) + \frac{2}{R_2 C} \qquad (3.5)$$

$$G = \mu = 1 + \frac{R_4}{R_3}$$

where μ is the gain of the VCVS, as well as the gain of the filter. Higher even-order high-pass filters may be obtained by factoring the transfer function of (3.2) into second-order factors like (3.4), realizing each factor with a stage like Fig. 3.4, and cascading the stages.

The procedure for constructing a second-order high-pass filter or a second-order stage of a higher-order high-pass filter is exactly like that of a low-pass filter described in Sec. 2.5 of Chapter 2. The procedure is summarized in Sec. 3.7, where the general circuit is shown followed by the design tables, which for the VCVS high-pass case are Tables 3-2 through 3-25. The K parameter for use with the tables may be calculated, as in the low-pass case, from

$$K = \frac{100}{f_c C'} \qquad (3.6)$$

where C' is the value of C in microfarads. Alternatively, K may be read from the appropriate one of Figs. 3-15a, b, or c.

As an example, suppose we want a fourth-order, 1/2 dB, high-pass Chebyshev filter with $f_c = 500$ Hz and a gain of 2. If we select capacitances of $C = 0.1 \, \mu$F, then by Eq. (3.6), the K parameter is 2. The resistances are given in Table 3-10 for $K = 1$, and hence for our case we must multiply these values by 2. The results in kilohms are for stage 1, $R_1 = 1.80$, $R_2 = 5.99$, $R_3 = 20.45$, $R_4 = 8.47$, and for stage 2, $R_1 = 1.77$, $R_2 = 2.04$, $R_3 = 6.97$, and $R_4 = 2.89$. We then construct the two stages in accordance with Fig. 3-4,

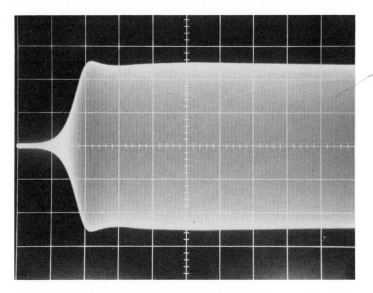

Figure 3-5.   A fourth-order, 1/2 dB, high-pass Chebyshev response.

using resistances as close as possible to those obtained from the table. This circuit was constructed with 5% tolerance resistors with values of 1.8, 6.2, 20, 8.2, 1.8, 2.0, 6.8 and 3.0 kΩ respectively, resulting in f_c = 496 Hz, a gain of 2.04, and a ripple very near 0.5. The picture of the response is shown in Fig. 3-5 with a horizontal scale of 500 Hz/division.

The advantages of the VCVS high-pass filter are identical to those of the VCVS low-pass filter given in Sec. 2.5.

3.3   Infinite-Gain Multiple-Feedback High-Pass Filters

A circuit analogous to the infinite-gain multiple-feedback low-pass filter of Sec. 2.6 is the *infinite-gain multiple-feedback high-pass filter* (infinite-gain MFB) of Fig. 3-6. It has the advantage of one fewer passive element, but it

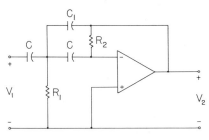

Figure 3-6.   A second-order infinite-gain multiple-feedback high-pass filter.

cannot be tuned to the proper gain as readily as the VCVS filter of the previous section.

Analysis of Fig. 3-6 shows that it achieves Eq. (3.4) with

$$a_0 = \frac{1}{R_1 R_2 C C_1}$$

$$a_1 = \frac{1}{R_2 C C_1} (2C + C_1) \tag{3.7}$$

$$G = -\frac{C}{C_1}$$

Thus the infinite-gain MFB high-pass filter has an inverting gain, as was the case for its low-pass counterpart.

The relative merits of the infinite-gain MFB high-pass filter are identical to those of its low-pass counterpart of Sec. 2.6.

A summary of the design procedure for the infinite-gain MFB high-pass filter is given in Sec. 3.8. The general circuit, practical suggestions, and the appropriate design tables, Tables 3-26 through 3-49, are included.

3.4   Biquad High-Pass Filters

The low-pass biquad circuit of Sec. 2.7 has its analogy for the high-pass case in the biquad circuit of Fig. 3-7 [33]. As in the low-pass case, the biquad requires more elements but this disadvantage is offset by its excellent tuning

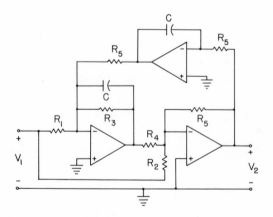

Figure 3-7.   A second order biquad high-pass filter.

features and stability. The biquad circuit achieves Eq. (3.4), for both C and R_5 normalized to 1 and an inverting gain $-G$ $(G > 0)$, with

$$a_0 = \frac{1}{R_4}$$

$$a_1 = \frac{1}{R_3} \tag{3.8}$$

$$G = \frac{1}{R_2} = \frac{a_0}{a_1 R_1}$$

The denormalization to obtain the desired cutoff and practical element values is identical to that described earlier in Sec. 2.4.

The advantages of the biquad high-pass filter are the same as those of the biquad low-pass filter, and are listed in Sec. 2.7.

The general procedure for a practical biquad high-pass filter design is given in Sec. 3.9. The pertinent design tables are Tables 3-50 through 3-53.

3.5 Multiple-Feedback High-Pass Filters

High-pass filters of our own design [31] are collected in this section for orders $n = 3$ to 8. These circuits are of the VCVS type with feedback paths provided through a set of resistors. Accordingly, we refer to the filters as *multiple-feedback high-pass filters*. With the capacitor C of the circuits and the cutoff frequency ω_c both normalized to 1, the transfer function (3.2) is achieved for various values of n with the equations given below. To obtain the normalized circuit element values, these equations were then iterated on a digital computer. Denormalization is then accomplished as described earlier in Sec. 2.4.

For $n = 3$ the circuit is shown in Fig. 3-8, and Eq. (3.2) is achieved with

$$a_0 = \frac{1}{R_1 R_2 R_3}$$

$$a_1 = \frac{2}{R_2 R_3} + \frac{A}{R_1} \tag{3.9}$$

$$a_2 = 2A + \frac{1}{R_1} - \frac{1}{R_3}$$

$$G = \mu$$

where

$$\mu = 1 + \frac{R_5}{R_4}$$

$$\tag{3.10}$$

$$A = \frac{2}{R_3} + \frac{1 - \mu}{R_2}$$

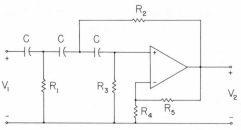

Figure 3-8. A third-order multiple-feedback high-pass filter.

For $n = 4$, we use Fig. 3-9, for which

$$a_0 = \frac{A}{R_3 R_4}$$

$$a_1 = \frac{B}{R_3 R_4} + AD$$

$$a_2 = BD + A + \frac{1}{R_3 R_4}$$ \hfill (3.11)

$$a_3 = B + D - \frac{\mu}{R_5}$$

$$G = \mu$$

where

$$\mu = 1 + \frac{R_7}{R_6}$$

$$A = \frac{1}{R_1}\left(\frac{1}{R_2} + \frac{1}{R_5}\right)$$ \hfill (3.12)

$$B = \frac{2}{R_1} + \frac{1}{R_5}$$

$$D = \frac{2}{R_3} + \frac{1 - \mu}{R_4}$$

For $n = 5$, we use Fig. 3-10, for which

$$a_0 = AB$$
$$a_1 = AE + BD$$
$$a_2 = A + DE + BF$$ \hfill (3.13)
$$a_3 = B + D + EF$$

$$a_4 = E + F - \frac{\mu}{R_6}$$

$$G = \mu$$

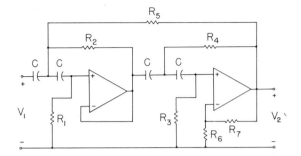

Figure 3-9. A fourth-order multiple-feedback high-pass filter.

where

$$\mu = 1 + \frac{R_8}{R_7}$$

$$A = \frac{1}{R_3 R_4 R_5}$$

$$B = \frac{1}{R_1} \left(\frac{1}{R_2} + \frac{1}{R_6} \right)$$

$$D = \frac{2}{R_4} \left(\frac{1}{R_3} + \frac{1}{R_5} \right) + \frac{1 - \mu}{R_3 R_5} \qquad (3.14)$$

$$E = \frac{2}{R_1} + \frac{1}{R_6}$$

$$F = \frac{1}{R_3} + \frac{3}{R_4} + \frac{2(1 - \mu)}{R_5}$$

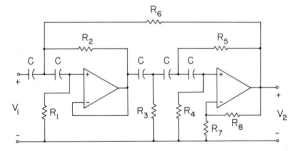

Figure 3-10. A fifth-order multiple-feedback high-pass filter.

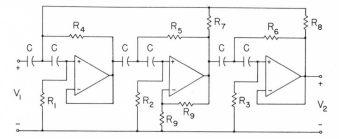

Figure 3-11. A sixth-order multiple-feedback high-pass filter.

For $n = 6$, we use Fig. 3-11, for which

$$a_0 = ABD$$

$$a_1 = AED + BDF + \frac{2AB}{R_3}$$

$$a_2 = AD + DEF + AB + BD + \frac{2(AE + BF)}{R_3}$$

$$a_3 = DF + DE + AE + BF + \frac{2(A + EF + B)}{R_3} - \frac{2D}{R_7} \qquad (3.15)$$

$$a_4 = A + D + B + EF + \frac{2\left(E + F - \dfrac{2}{R_7}\right)}{R_3}$$

$$a_5 = E + F + 2\left(\frac{1}{R_3} - \frac{1}{R_8} - \frac{1}{R_7}\right)$$

$$G = 2$$

where

$$A = \frac{1}{R_2 R_5}$$

$$B = \frac{1}{R_1}\left(\frac{1}{R_4} + \frac{1}{R_7} + \frac{1}{R_8}\right)$$

$$D = \frac{1}{R_3 R_6} \qquad (3.16)$$

$$E = \frac{2}{R_1} + \frac{1}{R_7} + \frac{1}{R_8}$$

$$F = \frac{2}{R_2} - \frac{1}{R_5}$$

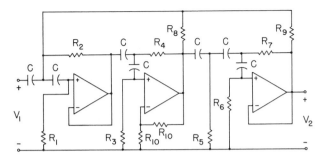

Figure 3-12. A seventh-order multiple-feedback high-pass filter.

For $n = 7$, we use Fig. 3-12, for which

$$a_0 = ABD$$
$$a_1 = ABL + D(BE + AF)$$
$$a_2 = D(A + B + EF) + L(AF + BE) + ABH$$

$$a_3 = D(E + F) + L(A + B + EF) + H(AF + BE) + AB - \frac{2D}{R_8} \quad (3.17)$$

$$a_4 = D + L(E + F) + H(A + B + EF) + BE + AF - \frac{2L}{R_8}$$

$$a_5 = L + H(E + F) + A + B + EF - \frac{2H}{R_8}$$

$$a_6 = H + E + F - 2\left(\frac{1}{R_8} + \frac{1}{R_9}\right)$$

$$G = 2$$

where

$$A = \frac{1}{R_1}\left(\frac{1}{R_2} + \frac{1}{R_8} + \frac{1}{R_9}\right)$$

$$B = \frac{1}{R_3 R_4}$$

$$D = \frac{1}{R_5 R_6 R_7}$$

$$E = \frac{2}{R_1} + \frac{1}{R_8} + \frac{1}{R_9} \quad (3.18)$$

$$F = \frac{2}{R_3} - \frac{1}{R_4}$$

$$H = \frac{3}{R_6} + \frac{1}{R_5}$$

$$L = \frac{2}{R_5 R_6} + \frac{2}{R_6 R_7}$$

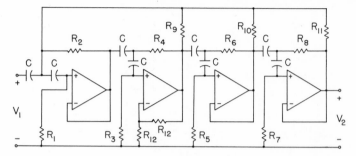

Figure 3-13.   An eighth-order multiple-feedback high-pass filter.

For $n = 8$, we use Fig. 3-13, for which

$$a_0 = ABDE$$
$$a_1 = AD(EF + BH) + BEL$$
$$a_2 = ADM + L(EF + BH) + BEN$$

$$a_3 = AD(F + H) + LM + N(BH + EF) + BEP - \frac{2AD}{R_9}$$

$$a_4 = AD + L(F + H) + MN + P(BH + EF) + BE - \frac{2L}{R_9} \quad (3.19)$$

$$a_5 = L + N(F + H) + MP + BH + EF - \frac{2N}{R_9} - \frac{2A}{R_{10}}$$

$$a_6 = N + P(F + H) + M - \frac{2P}{R_9} - \frac{4}{R_7 R_{10}}$$

$$a_7 = F + H + P - 2\left(\frac{1}{R_9} + \frac{1}{R_{10}} + \frac{1}{R_{11}}\right)$$

$$G = 2$$

where

$$A = \frac{1}{R_7 R_8}$$

$$B = \frac{1}{R_1}\left(\frac{1}{R_2} + \frac{1}{R_9} + \frac{1}{R_{10}} + \frac{1}{R_{11}}\right)$$

$$D = \frac{1}{R_5 R_6}$$

$$E = \frac{1}{R_3 R_4}$$

$$F = \frac{2}{R_1} + \frac{1}{R_9} + \frac{1}{R_{10}} + \frac{1}{R_{11}}$$

$$H = \frac{2}{R_3} - \frac{1}{R_4} \tag{3.20}$$

$$L = \frac{2A}{R_5} + \frac{2D}{R_7}$$

$$M = B + E + FH$$

$$N = A + D + \frac{4}{R_5 R_7}$$

$$P = 2\left(\frac{1}{R_5} + \frac{1}{R_7}\right)$$

As in the low-pass case, the multiple-feedback high-pass circuits present the mathematical difficulty of solving the various sets of equations for each order. The advantages are those of a multiple-feedback structure over a cascaded structure, summarized in Sec. 2.8.

The general design procedure for the multiple-feedback high-pass filters is given in Sec. 3.10. The pertinent tables for use in the design procedure are Tables 3-54 through 3-69. Examples of actual designs are given in the next section.

3.6   Examples of Multiple-Feedback High-Pass Filters

As an example of the procedure for obtaining multiple-feedback high-pass filters as described in the previous section, suppose we want a seventh-order, $\frac{1}{2}$ dB, Chebyshev high-pass filter with $f_c = 1000$ Hz and $G = 2$. Selecting C in Fig. 3-12 to be 0.01 μF, we have by Eq. (3.6) a K parameter of 10. Multiplying the resistances of Table 3-68 for the 0.5 dB case by 10 we have the resistances of the filter given by $R_1 = 28.97$, $R_2 = 4.07$, $R_3 = 33.48$, $R_4 = 12.85$, $R_5 = 3.31$, $R_6 = 26.40$, $R_7 = 3.35$, $R_8 = 20.87$, $R_9 = 44.95$, and $R_{10} = 67.00$ (all kilohms). The measured values used were, respectively, 28.7, 4.0, 33.6, 13.0, 3.32, 26.3, 3.32, 20.5, 45, and 67. The results were $f_c = 990$ Hz, $G = 1.96$, and a ripple of 0.53 dB. The response is shown in Fig. 3-14e.

For comparative purposes, other responses shown in Fig. 3-14 are (a) a third-order Butterworth with $f_c = 10,000$ Hz and $G = 1$, (b) a fourth-order, 2 dB Chebyshev with $f_c = 10,000$ Hz and $G = 2$, (c) a fifth-order, 0.5 dB Chebyshev with $f_c = 200$ Hz and $G = 2$, (d) a sixth-order Butterworth with $f_c = 1000$ Hz and $G = 2$, and (f) an eighth-order, 0.1 dB Chebyshev with $f_c = 1000$ Hz and $G = 2$.

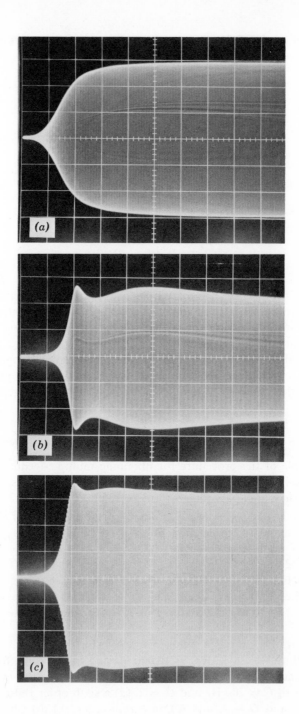

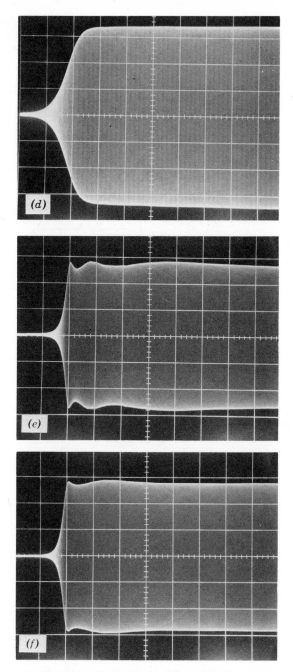

Figure 3-14. (*a*) A third-order Butterworth; (*b*) a fourth-order, 2 dB Chebyshev; (*c*) a fifth-order, 0.5 dB Chebyshev; (*d*) a sixth-order Butterworth; (*e*) a seventh-order, 0.5 dB Chebyshev; and (*f*) an eighth-order, 0.1 dB, Chebyshev high-pass response.

3.7   Summary of VCVS High-Pass Filter Design Procedure

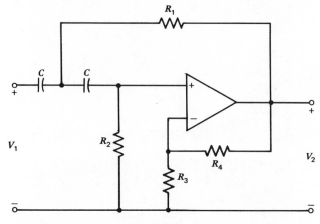

General circuit

Procedure

Given cutoff f_c (hertz), gain G, order n, and filter type (Butterworth or Chebyshev), perform the following steps for a second-order filter, or for each stage of a higher-order cascaded filter ($n = 4, 6, 8$).

1.   Select a value of capacitance C and determine a K parameter from

$$K = \frac{100}{f_c C'}$$

where C' is the value of C in microfarads. Alternately, K may be found from Fig. 3-15a, b, or c. For higher-order designs (say, $n > 4$), it is better to use the equation since greater accuracy is required.

2.   Find the remaining element values from the appropriate one of Tables 3-2 through 3-25. The resistances in the tables are given for $K = 1$, and hence their values must be multiplied by the K parameter of step 1 to yield the resistances of the circuit.

3.   Select standard resistance values that are as close as possible to those indicated by the table and construct the filter, or its stages, in accordance with the general circuit.

Comments and Suggestions

(a) In the case of multiple-stage filters ($n > 2$), the K parameter for each stage need not be the same. That is, one may select a different C for each stage and using the given f_c of the overall filter calculate a K parameter for each

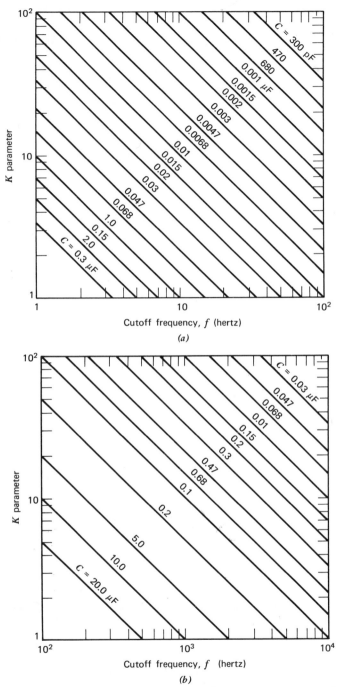

Figure 3-15. (*a*) *K* parameter versus frequency. (*b*) *K* parameter versus frequency.

89

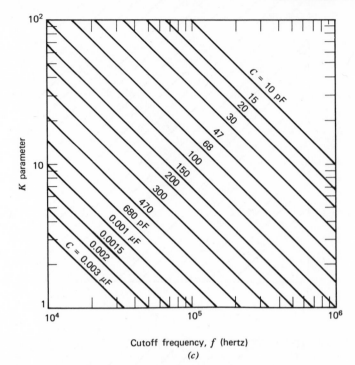

Cutoff frequency, f (hertz)
(c)

Figure 3-15. (c) K parameter versus frequency.

stage. The resistances for each stage are those of the table times the K parameter of the stage.

(b) For best performance, the input resistance of the op-amp should be at least 10 times $R_{eq} = R_2$. For a specific op-amp, this condition can generally be satisfied by proper selection of C to obtain a suitable K parameter.

(c) The values in the tables for R_3 and R_4 were determined to minimize the dc offset of the op-amp. Other values of R_3 and R_4 may be used as long as the ratio R_4/R_3 is the same as that of the table values.

(d) Standard resistance values of 5% tolerance normally yield acceptable results in the lower-order cases. For fifth and sixth orders, resistances of 2% tolerance probably should be used, and for seventh and eighth orders, 1% tolerance resistances probably should be used. In all cases, for best performance, resistance values close to those indicated by the tables should be used.

In the case of capacitors, percentage tolerances should parallel those given above for the resistors, for best results. Since precision capacitors are relatively expensive, it may be desirable to use capacitors of higher tolerances, in which case trimming is generally required. In the case of low orders ($n \leq 4$), 10% capacitors are quite often satisfactory.

(e) The gain of each stage of the filter is $1 + R_4/R_3$, which can be adjusted to the correct value by using a potentiometer in lieu of resistors R_3 and R_4. This is accomplished by connecting the center tap of the potentiometer to the inverting input of the op-amp. These gain adjustments are very useful in tuning the overall response of the filter.

(f) Finally, the open-loop gain of the op-amp should be at least 50 times the gain of the filter at f_a, and the desired peak-to-peak voltage at f_a should not exceed $10^6/\pi f_a$ times the slew rate of the op-amp, where f_a is the highest frequency desired in the passband. Thus, for high values of f_a, externally compensated op-amps may be required.

A specific example of a fourth-order VCVS design was given in Sec. 3.2.

Table 3-2. Second-Order High-Pass Butterworth VCVS Filter Designs

Gain	Circuit Element Values[a]					
	1	2	4	6	8	10
R_1	1.125	1.821	2.592	3.141	3.593	3.985
R_2	2.251	1.391	0.977	0.806	0.705	0.636
R_3	Open	2.782	1.303	0.968	0.806	0.706
R_4	0	2.782	3.910	4.838	5.640	6.356

[a] Resistances in kilohms for a K parameter of 1.

Table 3-3. Second-Order High-Pass Chebyshev VCVS Filter Designs (0.1 dB)

Gain	Circuit Element Values[a]					
	1	2	4	6	8	10
R_1	1.888	3.199	4.615	5.621	6.445	7.161
R_2	4.446	2.623	1.818	1.493	1.302	1.172
R_3	Open	5.247	2.425	1.792	1.488	1.302
R_4	0	5.247	7.274	8.959	10.417	11.719

[a] Resistances in kilohms for a K parameter of 1.

Table 3-4. Second-Order High-Pass Chebyshev VCVS
Filter Designs (0.5 dB)

	Circuit Element Values[a]					
Gain	1	2	4	6	8	10
R_1	1.134	2.065	3.034	3.717	4.277	4.763
R_2	3.385	1.860	1.266	1.033	0.898	0.806
R_3	Open	3.720	1.688	1.240	1.026	0.896
R_4	0	3.720	5.064	6.199	7.183	8.063

[a] Resistances in kilohms for a K parameter of 1.

Table 3-5. Second-Order High-Pass Chebyshev VCVS
Filter Designs (1 dB)

	Circuit Element Values[a]					
Gain	1	2	4	6	8	10
R_1	0.874	1.697	2.530	3.115	3.594	4.009
R_2	3.197	1.646	1.104	0.897	0.777	0.697
R_3	Open	3.292	1.472	1.076	0.888	0.774
R_4	0	3.292	4.416	5.379	6.217	6.967

[a] Resistances in kilohms for a K parameter of 1.

Table 3-6. Second-Order High-Pass Chebyshev VCVS
Filter Designs (2 dB)

	Circuit Element Values[a]					
Gain	1	2	4	6	8	10
R_1	0.640	1.390	2.117	2.625	3.040	3.399
R_2	3.259	1.500	0.985	0.794	0.686	0.613
R_3	Open	3.000	1.313	0.953	0.784	0.681
R_4	0	3.000	3.939	4.765	5.486	6.133

[a] Resistances in kilohms for a K parameter of 1.

Table 3-7. Second-order High-Pass Chebyshev VCVS Filter Designs (3 dB)

Gain	Circuit Element Values[a]					
	1	2	4	6	8	10
R_1	0.513	1.238	1.917	2.389	2.775	3.109
R_2	3.494	1.449	0.936	0.750	0.646	0.577
R_3	Open	2.898	1.248	0.901	0.739	0.641
R_4	0	2.898	3.743	4.503	5.170	5.768

[a] Resistances in kilohms for a K parameter of 1.

Table 3-8. Fourth-Order High-Pass Butterworth Cascaded VCVS Filter Designs

Gain	Circuit Element Values[a]						Stage
	1	2	6	10	36	100	
R_1	0.609	1.090	1.693	1.987	2.839	3.694	
R_2	4.159	2.323	1.496	1.275	0.892	0.686	
R_3	Open	7.933	2.528	1.864	1.071	0.762	1
R_4	0	3.286	3.664	4.031	5.353	6.856	
R_1	1.470	1.767	2.277	2.546	3.357	4.191	
R_2	1.723	1.433	1.113	0.995	0.755	0.604	
R_3	Open	4.894	1.880	1.455	0.905	0.672	2
R_4	0	2.027	2.725	3.146	4.528	6.045	

[a] Resistances in kilohms for a K parameter of 1.

Table 3-9. Fourth-Order High-Pass Chebyshev Cascaded VCVS Filter
Designs (0.1 dB)

	Circuit Element Values[a]						
Gain	1	2	6	10	36	100	Stage
R_1	0.420	1.071	1.787	2.130	3.120	4.109	
R_2	8.013	3.144	1.885	1.581	1.080	0.820	
R_3	Open	10.734	3.186	2.313	1.296	0.911	1
R_4	0	4.446	4.618	5.001	6.478	8.197	
R_1	1.015	1.272	1.691	1.909	2.557	3.220	
R_2	1.554	1.240	0.933	0.827	0.617	0.490	
R_3	Open	4.235	1.577	1.209	0.740	0.544	2
R_4	0	1.754	2.285	2.614	3.702	4.900	

[a] Resistances in kilohms for a K parameter of 1.

Table 3-10. Fourth-Order High-Pass Chebyshev Cascaded VCVS Filter
Designs (0.5 dB)

	Circuit Element Values[a]						
Gain	1	2	6	10	36	100	Stage
R_1	0.279	0.899	1.544	1.852	2.738	3.624	
R_2	9.653	2.995	1.745	1.455	0.984	0.743	
R_3	Open	10.227	2.949	2.127	1.180	0.826	1
R_4	0	4.236	4.274	4.600	5.902	7.433	
R_1	0.674	0.885	1.213	1.381	1.877	2.380	
R_2	1.340	1.020	0.744	0.654	0.481	0.379	
R_3	Open	3.483	1.258	0.956	0.577	0.421	2
R_4	0	1.443	1.823	2.068	2.887	3.793	

[a] Resistances in kilohms for a K parameter of 1.

Table 3-11. Fourth-Order High-Pass Chebyshev Cascaded VCVS Filter Designs (1 dB)

Gain	1	2	6	10	36	100	Stage
			Circuit Element Values[a]				
R_1	0.222	0.839	1.461	1.758	2.613	3.466	
R_2	11.252	2.979	1.710	1.421	0.956	0.721	
R_3	Open	10.170	2.890	2.078	1.148	0.801	1
R_4	0	4.212	4.189	4.494	5.738	7.209	
R_1	0.536	0.735	1.033	1.183	1.625	2.073	
R_2	1.320	0.962	0.685	0.598	0.436	0.341	
R_3	Open	3.286	1.158	0.875	0.523	0.379	2
R_4	0	1.361	1.679	1.892	2.613	3.415	

[a] Resistances in kilohms for a K parameter of 1.

Table 3-12. Fourth-Order High-Pass Chebyshev Cascaded VCVS Filter Designs (2 dB)

Gain	1	2	6	10	36	100	Stage
			Circuit Element Values[a]				
R_1	0.167	0.786	1.392	1.680	2.510	3.338	
R_2	14.092	2.991	1.690	1.400	0.937	0.705	
R_3	Open	10.213	2.856	2.047	1.125	0.783	1
R_4	0	4.230	4.140	4.427	5.623	7.047	
R_1	0.403	0.598	0.870	1.006	1.403	1.803	
R_2	1.393	0.939	0.645	0.558	0.400	0.311	
R_3	Open	3.207	1.090	0.816	0.480	0.346	2
R_4	0	1.328	1.580	1.764	2.400	3.112	

[a] Resistances in kilohms for a K parameter of 1.

Table 3-13. Fourth-Order High-Pass Chebyshev Cascaded VCVS Filter Designs (3 dB)

	Circuit Element Values[a]						
Gain	1	2	6	10	36	100	Stage
R_1	0.136	0.759	1.357	1.642	2.460	3.277	
R_2	16.876	3.012	1.686	1.393	0.930	0.698	
R_3	Open	10.285	2.848	2.038	1.116	0.776	1
R_4	0	4.260	4.129	4.406	5.579	6.981	
R_1	0.327	0.524	0.785	0.914	1.290	1.667	
R_2	1.517	0.948	0.632	0.543	0.385	0.298	
R_3	Open	3.237	1.068	0.794	0.462	0.331	2
R_4	0	1.341	1.548	1.717	2.310	2.978	

[a] Resistances in kilohms for a K parameter of 1.

Table 3-14. Sixth-Order High-Pass Butterworth Cascaded VCVS Filter Designs

	Circuit Element Values[a]						
Gain	1	4	10	50	100	500	Stage
R_1	0.412	1.093	1.433	2.061	2.363	3.177	
R_2	6.149	2.318	1.768	1.229	1.072	0.797	
R_3	Open	6.264	3.300	1.687	1.366	0.912	1
R_4	0	3.680	3.809	4.527	4.975	6.328	
R_1	1.125	1.593	1.896	2.490	2.783	3.580	
R_2	2.251	1.591	1.336	1.017	0.910	0.708	
R_3	Open	4.298	2.493	1.396	1.160	0.810	2
R_4	0	2.525	2.878	3.747	4.225	5.616	
R_1	1.537	1.924	2.201	2.766	3.050	3.831	
R_2	1.648	1.317	1.151	0.916	0.831	0.661	
R_3	Open	3.558	2.147	1.257	1.059	0.757	3
R_4	0	2.090	2.479	3.373	3.855	5.248	

[a] Resistances in kilohms for a K parameter of 1.

Table 3-15. Sixth-Order High-Pass Chebyshev Cascaded VCVS Filter Designs (0.1 dB)

	Circuit Element Values[a]						
Gain	1	4	10	50	100	500	Stage
R_1	0.183	1.012	1.380	2.053	2.376	3.243	
R_2	15.673	2.826	2.074	1.394	1.204	0.882	1
R_3	Open	7.637	3.870	1.913	1.535	1.009	
R_4	0	4.486	4.468	5.134	5.590	7.003	
R_1	0.499	1.011	1.289	1.808	2.059	2.736	
R_2	3.537	1.745	1.369	0.976	0.857	0.645	2
R_3	Open	4.715	2.555	1.339	1.092	0.738	
R_4	0	2.770	2.949	3.594	3.977	5.119	
R_1	0.681	0.899	1.049	1.346	1.494	1.900	
R_2	0.979	0.742	0.636	0.496	0.446	0.351	3
R_3	Open	2.005	1.187	0.680	0.569	0.402	
R_4	0	1.178	1.371	1.826	2.072	2.788	

[a] Resistances in kilohms for a K parameter of 1.

Table 3-16. Sixth-Order High-Pass Chebyshev Cascaded VCVS Filter Designs (0.5 dB)

	Circuit Element Values[a]						
Gain	1	4	10	50	100	500	Stage
R_1	0.124	0.936	1.286	1.928	2.235	3.060	
R_2	20.968	2.767	2.014	1.344	1.160	0.847	1
R_3	Open	7.479	3.760	1.845	1.478	0.969	
R_4	0	4.393	4.340	4.942	5.382	6.720	
R_1	0.338	0.853	1.113	1.595	1.827	2.452	
R_2	4.426	1.753	1.343	0.937	0.818	0.610	2
R_3	Open	4.738	2.506	1.286	1.043	0.697	
R_4	0	2.783	2.893	3.452	3.797	4.838	
R_1	0.461	0.643	0.762	0.997	1.112	1.427	
R_2	0.862	0.619	0.522	0.399	0.358	0.279	3
R_3	Open	1.672	0.974	0.548	0.456	0.319	
R_4	0	0.982	1.124	1.470	1.660	2.211	

[a] Resistances in kilohms for a K parameter of 1.

Table 3-17. Sixth-Order High-Pass Chebyshev Cascaded VCVS Filter Designs (1 dB)

	Circuit Element Values[a]						
Gain	1	4	10	50	100	500	Stage
R_1	0.099	0.909	1.254	1.885	2.188	3.000	
R_2	25.358	2.759	2.001	1.331	1.147	0.836	
R_3	Open	7.457	3.735	1.827	1.462	0.957	1
R_4	0	4.380	4.311	4.904	5.325	6.639	
R_1	0.270	0.793	1.048	1.519	1.745	2.353	
R_2	5.225	1.781	1.348	0.930	0.810	0.600	
R_3	Open	4.812	2.515	1.277	1.032	0.687	2
R_4	0	2.827	2.903	3.427	3.758	4.765	
R_1	0.369	0.541	0.650	0.861	0.965	1.248	
R_2	0.855	0.584	0.486	0.367	0.327	0.253	
R_3	Open	1.578	0.907	0.503	0.417	0.290	3
R_4	0	0.927	1.047	1.351	1.519	2.010	

[a] Resistances in kilohms for a K parameter of 1.

Table 3-18. Sixth-Order High-Pass Chebyshev Cascaded VCVS Filter Designs (2 dB)

	Circuit Element Values[a]						
Gain	1	4	10	50	100	500	Stage
R_1	0.075	0.886	1.226	1.850	2.148	2.951	
R_2	32.728	2.762	1.995	1.323	1.139	0.829	
R_3	Open	7.464	3.723	1.816	1.452	0.949	1
R_4	0	4.384	4.298	4.873	5.286	6.581	
R_1	0.204	0.740	0.991	1.452	1.673	2.268	
R_2	6.609	1.824	1.363	0.930	0.807	0.595	
R_3	Open	4.930	2.543	1.276	1.028	0.681	2
R_4	0	2.896	2.936	3.425	3.745	4.723	
R_1	0.279	0.446	0.546	0.739	0.833	1.087	
R_2	0.907	0.568	0.463	0.343	0.304	0.233	
R_3	Open	1.534	0.865	0.470	0.388	0.266	3
R_4	0	0.901	0.998	1.262	1.411	1.849	

[a] Resistances in kilohms for a K parameter of 1.

Table 3-19. Sixth-Order High-Pass Chebyshev Cascaded VCVS Filter Designs (3 dB)

Gain	Circuit Element Values[a]						Stage
	1	4	10	50	100	500	
R_1	0.061	0.874	1.212	1.832	2.129	2.927	
R_2	39.751	2.768	1.995	1.320	1.136	0.826	1
R_3	Open	7.480	3.723	1.812	1.448	0.945	
R_4	0	4.394	4.298	4.863	5.273	6.559	
R_1	0.166	0.712	0.961	1.418	1.637	2.226	
R_2	7.952	1.857	1.376	0.932	0.808	0.594	2
R_3	Open	5.019	2.568	1.280	1.029	0.679	
R_4	0	2.948	2.965	3.435	3.749	4.713	
R_1	0.227	0.395	0.491	0.675	0.764	1.004	
R_2	0.991	0.570	0.458	0.333	0.295	0.224	3
R_3	Open	1.541	0.854	0.458	0.376	0.256	
R_4	0	0.905	0.986	1.228	1.367	1.778	

[a] Resistances in kilohms for a K parameter of 1.

Table 3-20.   Eighth-Order High-Pass Butterworth Cascaded VCVS
Filter Designs

Gain	1	4	10	50	100	500	Stage
			Circuit Element Values[a]				
R_1	0.310	0.896	1.160	1.613	1.817	2.334	
R_2	8.158	2.827	2.183	1.570	1.394	1.085	1
R_3	Open	9.653	4.989	2.517	2.038	1.376	
R_4	0	3.998	3.883	4.176	4.408	5.132	
R_1	0.884	1.291	1.529	1.958	2.155	2.660	
R_2	2.865	1.963	1.657	1.294	1.175	0.952	2
R_3	Open	6.701	3.785	2.074	1.719	1.208	
R_4	0	2.775	2.946	3.441	3.717	4.503	
R_1	1.323	1.643	1.855	2.255	2.444	2.933	
R_2	1.914	1.542	1.366	1.123	1.036	0.864	3
R_3	Open	5.265	3.120	1.800	1.516	1.095	
R_4	0	2.181	2.429	2.987	3.278	4.083	
R_1	1.561	1.845	2.043	2.427	2.610	3.090	
R_2	1.623	1.373	1.240	1.044	0.970	0.820	4
R_3	Open	4.687	2.832	1.673	1.419	1.040	
R_4	0	1.941	2.204	2.775	3.069	3.877	

[a] Resistances in kilohms for a K parameter of 1.

Table 3-21.   Eighth-Order High-Pass Chebyshev Cascaded VCVS Filter Designs (0.1 dB)

Gain	Circuit Element Values[a]						Stage
	1	4	10	50	100	500	
R_1	0.102	0.802	1.079	1.551	1.763	2.299	
R_2	26.606	3.379	2.511	1.747	1.537	1.179	1
R_3	Open	11.539	5.737	2.800	2.247	1.495	
R_4	0	4.779	4.465	4.645	4.859	5.573	
R_1	0.290	0.808	1.044	1.449	1.631	2.093	
R_2	6.979	2.503	1.938	1.397	1.241	0.967	2
R_3	Open	8.547	4.428	2.239	1.814	1.226	
R_4	0	3.540	3.447	3.715	3.923	4.573	
R_1	0.434	0.732	0.893	1.177	1.307	1.636	
R_2	2.429	1.440	1.180	0.896	0.807	0.645	3
R_3	Open	4.917	2.697	1.436	1.180	0.817	
R_4	0	2.036	2.099	2.382	2.552	3.048	
R_1	0.512	0.633	0.713	0.866	0.937	1.124	
R_2	0.721	0.583	0.517	0.426	0.394	0.328	4
R_3	Open	1.991	1.182	0.683	0.576	0.416	
R_4	0	0.825	0.920	1.133	1.244	1.552	

[a] Resistances in kilohms for a K parameter of 1.

Table 3-22.   Eighth-Order High-Pass Chebyshev Cascaded VCVS Filter
Designs (0.5 dB)

Gain	Circuit Element Values[a]						Stage
	1	4	10	50	100	500	
R_1	0.069	0.764	1.034	1.493	1.700	2.221	
R_2	36.922	3.354	2.479	1.716	1.508	1.154	
R_3	Open	11.453	5.664	2.751	2.205	1.464	1
R_4	0	4.744	4.408	4.564	4.769	5.457	
R_1	0.198	0.730	0.959	1.351	1.527	1.973	
R_2	9.498	2.571	1.957	1.390	1.230	0.952	
R_3	Open	8.780	4.472	2.228	1.798	1.207	2
R_4	0	3.637	3.481	3.696	3.888	4.502	
R_1	0.296	0.606	0.761	1.029	1.150	1.458	
R_2	3.070	1.499	1.194	0.883	0.790	0.623	
R_3	Open	5.117	2.729	1.416	1.155	0.790	3
R_4	0	2.119	2.124	2.349	2.498	2.947	
R_1	0.349	0.451	0.517	0.639	0.696	0.843	
R_2	0.639	0.494	0.431	0.349	0.321	0.265	
R_3	Open	1.687	0.986	0.560	0.469	0.336	4
R_4	0	0.699	0.767	0.929	1.014	1.252	

[a] Resistances in kilohms for a K parameter of 1.

Table 3-23.   Eighth-Order High-Pass Chebyshev Cascaded VCVS Filter Design (1 dB)

Gain	Circuit Element Values[a]						Stage
	1	4	10	50	100	500	
R_1	0.056	0.751	1.018	1.473	1.678	2.195	
R_2	45.196	3.355	2.473	1.709	1.501	1.147	
R_3	Open	11.455	5.651	2.739	2.195	1.455	1
R_4	0	4.745	4.398	4.545	4.745	5.425	
R_1	0.159	0.701	0.928	1.315	1.489	1.930	
R_2	11.551	2.616	1.976	1.394	1.231	0.950	
R_3	Open	8.933	4.515	2.234	1.800	1.205	2
R_4	0	3.700	3.514	3.706	3.892	4.492	
R_1	0.237	0.558	0.710	0.973	1.092	1.393	
R_2	3.636	1.547	1.215	0.887	0.791	0.620	
R_3	Open	5.284	2.777	1.422	1.156	0.786	3
R_4	0	2.188	2.161	2.359	2.500	2.931	
R_1	0.280	0.378	0.438	0.549	0.601	0.733	
R_2	0.635	0.471	0.406	0.324	0.296	0.243	
R_3	Open	1.609	0.928	0.520	0.433	0.308	4
R_4	0	0.666	0.722	0.862	0.937	1.148	

[a] Resistances in kilohms for a K parameter of 1.

Table 3-24. Eighth-Order High-Pass Chebyshev Cascaded VCVS Filter
Designs (2 dB)

Gain	Circuit Element Values[a]						Stage
	1	4	10	50	100	500	
R_1	0.042	0.739	1.004	1.457	1.660	2.173	
R_2	58.897	3.363	2.473	1.705	1.496	1.143	1
R_3	Open	11.481	5.649	2.733	2.188	1.449	
R_4	0	4.755	4.397	4.534	4.731	5.404	
R_1	0.120	0.673	0.899	1.283	1.456	1.892	
R_2	14.973	2.671	2.001	1.402	1.235	0.950	2
R_3	Open	9.119	4.571	2.246	1.806	1.205	
R_4	0	3.777	3.558	3.727	3.906	4.494	
R_1	0.180	0.514	0.665	0.924	1.041	1.336	
R_2	4.611	1.613	1.246	0.897	0.796	0.620	3
R_3	Open	5.507	2.848	1.438	1.164	0.787	
R_4	0	2.281	2.217	2.385	2.518	2.933	
R_1	0.212	0.308	0.365	0.466	0.513	0.633	
R_2	0.675	0.464	0.392	0.307	0.279	0.226	4
R_3	Open	1.586	0.897	0.492	0.408	0.287	
R_4	0	0.657	0.698	0.816	0.882	1.069	

[a] Resistances in kilohms for a K parameter of 1.

Table 3-25. Eighth-Order High-Pass Chebyshev Cascaded VCVS Filter
Designs (3 dB)

Gain	Circuit Element Values[a]						Stage
	1	4	10	50	100	500	
R_1	0.034	0.732	0.997	1.448	1.651	2.162	
R_2	71.852	3.370	2.474	1.704	1.495	1.141	1
R_3	Open	11.506	5.654	2.731	2.186	1.447	
R_4	0	4.766	4.400	4.531	4.728	5.397	
R_1	0.098	0.658	0.883	1.266	1.438	1.872	
R_2	18.223	2.707	2.018	1.408	1.239	0.952	2
R_3	Open	9.242	4.611	2.257	1.813	1.207	
R_4	0	3.828	3.589	3.744	3.920	4.501	
R_1	0.146	0.490	0.640	0.898	1.013	1.306	
R_2	5.553	1.659	1.269	0.906	0.802	0.622	3
R_3	Open	5.664	2.900	1.451	1.173	0.789	
R_4	0	2.346	2.257	2.408	2.536	2.942	
R_1	0.173	0.270	0.325	0.423	0.467	0.581	
R_2	0.738	0.471	0.392	0.301	0.273	0.219	4
R_3	Open	1.609	0.895	0.483	0.399	0.278	
R_4	0	0.667	0.697	0.801	0.862	1.036	

[a] Resistances in kilohms for a K parameter of 1.

3.8   Summary of Infinite-Gain MFB High-Pass Filter Design Procedure

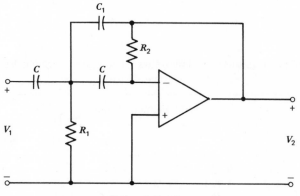

General circuit

Procedure

Given cutoff f_c (hertz), gain G, order n, and filter type (Butterworth or Chebyshev), perform the following steps for a second-order filter, or for each stage of a higher-order cascaded filter ($n = 4, 6, 8$).

1. Select a value of capacitance C and determine a K parameter from

$$K = \frac{100}{f_c C'}$$

where C' is the value of C in microfarads. Alternatively, K may be found from Fig. 3-15 *a*, *b*, or *c*. For higher-order designs (say, $n > 4$), it is better to use the equation since greater accuracy is required.

2. Find the remaining element values from the appropriate one of Tables 3-26 through 3-49 as follows. The values of C_1 are determined directly from the tables using the chosen value of C. The resistances in the tables are given for $K = 1$ and hence their values must be multiplied by the K parameter of step 1 to yield the resistances of the circuit.

3. Select standard resistance values that are as close as possible to those indicated by the table and construct the filter, or its stages, in accordance with the general circuit. In case C_1 is a multiple of C such as 0.47, and so forth, standard values of C_1 result if C is chosen as a power of 10 (i.e., 0.1, 1, 10, etc.) μF.

Comments and Suggestions

The comments and suggestions for the VCVS high-pass filter given in Sec. 3.7 apply as follows:

(a) Paragraphs (a), (b), (d), and (f) are directly applicable.
(b) Paragraphs (c) and (e) do not apply.

In addition, the following applies:

(c) The inverting gain of each stage of the filter is C/C_1. Gain adjustments can be made by trimming either C or C_1.
(d) For minimum dc offset, a resistance equal to R_2 can be placed in the noninverting input to ground.

The infinite-gain MFB high-pass filters were discussed in Sec. 3.3.

Table 3-26. Second-Order High-Pass Butterworth Infinite-Gain MFB Filter Designs

	Circuit Element Values[a]			
Gain	1	2	5	10
R_1	0.750	0.900	1.023	1.072
R_2	3.376	5.627	12.379	23.634
C_1	C	0.5C	0.2C	0.1C

[a] Resistances in kilohms for a K parameter of 1.

Table 3-27. Second-Order High-Pass Chebyshev Infinite-Gain MFB Filter Designs (0.1 dB)

	Circuit Element Values[a]			
Gain	1	2	5	10
R_1	1.258	1.510	1.716	1.798
R_2	6.669	11.115	24.453	46.684
C_1	C	0.5C	0.2C	0.1C

[a] Resistances in kilohms for a K parameter of 1.

Table 3-28. Second-Order High-Pass Chebyshev Infinite-Gain MFB Filter Designs (0.5 dB)

Gain	Circuit Element Values[a]			
	1	2	5	10
R_1	0.756	0.908	1.031	1.080
R_2	5.078	8.463	18.619	35.546
C_1	C	0.5C	0.2C	0.1C

[a] Resistances in kilohms for a K parameter of 1.

Table 3-29. Second-Order High-Pass Chebyshev Infinite-Gain MFB Filter Designs (1 dB)

Gain	Circuit Element Values[a]			
	1	2	5	10
R_1	0.582	0.699	0.794	0.832
R_2	4.795	7.992	17.583	33.568
C_1	C	0.5C	0.2C	0.1C

[a] Resistances in kilohms for a K parameter of 1.

Table 3-30. Second-Order High-Pass Chebyshev Infinite-Gain MFB Filter Designs (2 dB)

Gain	Circuit Element Values[a]			
	1	2	5	10
R_1	0.426	0.512	0.581	0.609
R_2	4.889	8.148	17.925	34.221
C_1	C	0.5C	0.2C	0.1C

[a] Resistances in kilohms for a K parameter of 1.

Table 3-31. Second-Order High-Pass Chebyshev Infinite-Gain MFB Filter Designs (3 dB)

Gain	Circuit Element Values[a]			
	1	2	5	10
R_1	0.342	0.411	0.467	0.489
R_2	5.241	8.736	19.219	36.690
C_1	C	0.5C	0.2C	0.1C

[a] Resistances in kilohms for a K parameter of 1.

Table 3-32. Fourth-Order High-Pass Butterworth Cascaded MFB Filter Designs

Gain	Circuit Element Values[a]				Stage
	1	4	25	100	
R_1	0.406	0.487	0.554	0.580	
R_2	6.238	10.397	22.874	43.669	1
C_1	C	0.5C	0.2C	0.1C	
R_1	0.980	1.176	1.337	1.400	
R_2	2.584	4.307	9.475	18.088	2
C_1	C	0.5C	0.2C	0.1C	

[a] Resistances in kilohms for a K parameter of 1.

Table 3-33. Fourth-Order High-Pass Chebyshev Cascaded MFB Filter Designs (0.1 dB)

Gain	Circuit Element Values[a]				Stage
	1	4	25	100	
R_1	0.280	0.336	0.382	0.400	
R_2	12.019	20.032	44.070	84.134	1
C_1	C	0.5C	0.2C	0.1C	
R_1	0.677	0.812	0.923	0.967	
R_2	2.332	3.886	8.550	16.322	2
C_1	C	0.5C	0.2C	0.1C	

[a] Resistances in kilohms for a K parameter of 1.

Table 3-34. Fourth-Order High-Pass Chebyshev
Cascaded MFB Filter Designs (0.5 dB)

Gain	Circuit Element Values[a]				Stage
	1	4	25	100	
R_1	0.186	0.223	0.254	0.266	
R_2	14.479	24.132	53.090	101.354	1
C_1	C	0.5C	0.2C	0.1C	
R_1	0.449	0.539	0.613	0.642	
R_2	2.010	3.350	7.370	14.069	2
C_1	C	0.5C	0.2C	0.1C	

[a] Resistances in kilohms for a K parameter of 1.

Table 3-35. Fourth-Order High-Pass Chebyshev
Cascaded MFB Filter Designs (1 dB)

Gain	Circuit Element Values[a]				Stage
	1	4	25	100	
R_1	0.148	0.178	0.202	0.212	
R_2	16.878	28.130	61.886	118.147	1
C_1	C	0.5C	0.2C	0.1C	
R_1	0.357	0.429	0.487	0.511	
R_2	1.980	3.300	7.260	13.860	2
C_1	C	0.5C	0.2C	0.1C	

[a] Resistances in kilohms for a K parameter of 1.

Table 3-36. Fourth-Order High-Pass Chebyshev
Cascaded MFB Filter Designs (2 dB)

Gain	Circuit Element Values[a]				Stage
	1	4	25	100	
R_1	0.111	0.134	0.152	0.159	
R_2	21.137	35.229	77.504	147.962	1
C_1	C	0.5C	0.2C	0.1C	
R_1	0.269	0.322	0.366	0.384	
R_2	2.089	3.482	7.659	14.622	2
C_1	C	0.5C	0.2C	0.1C	

[a] Resistances in kilohms for a K parameter of 1.

Table 3-37. Fourth-Order High-Pass Chebyshev
Cascaded MFB Filter Designs (3 dB)

Gain	Circuit Element Values[a]				Stage
	1	4	25	100	
R_1	0.090	0.108	0.123	0.129	
R_2	25.314	42.189	92.816	177.195	1
C_1	C	0.5C	0.2C	0.1C	
R_1	0.218	0.262	0.298	0.312	
R_2	2.275	3.792	8.343	15.928	2
C_1	C	0.5C	0.2C	0.1C	

[a] Resistances in kilohms for a K parameter of 1.

Table 3-38. Sixth-Order High-Pass Butterworth
Cascaded MFB Filter Designs

Gain	Circuit Element Values[a]				Stage
	1	4	25	100	
R_1	0.275	0.275	0.275	0.366	
R_2	9.224	9.224	9.224	27.672	1
C_1	C	C	C	0.25C	
R_1	0.750	0.900	1.023	1.023	
R_2	3.376	5.627	12.379	12.379	2
C_1	C	0.5C	0.2C	0.2C	
R_1	1.025	1.230	1.398	1.398	
R_2	2.472	4.119	9.062	9.062	3
C_1	C	0.5C	0.2C	0.2C	

[a] Resistances in kilohms for a K parameter of 1.

Table 3-39. Sixth-Order High-Pass Chebyshev
Cascaded MFB Filter Designs (0.1 dB)

Gain	Circuit Element Values[a]				Stage
	1	4	25	100	
R_1	0.122	0.122	0.122	0.162	
R_2	23.509	23.509	23.509	70.526	1
C_1	C	C	C	0.25C	
R_1	0.332	0.399	0.453	0.453	
R_2	5.306	8.843	19.455	19.455	2
C_1	C	0.5C	0.2C	0.2C	
R_1	0.454	0.545	0.619	0.619	
R_2	1.469	2.448	5.386	5.386	3
C_1	C	0.5C	0.2C	0.2C	

[a] Resistances in kilohms for a K parameter of 1.

Table 3-40. Sixth-Order High-Pass Chebyshev
Cascaded MFB Filter Designs (0.5 dB)

	Circuit Element Values[a]				
Gain	1	4	25	100	Stage
R_1	0.082	0.082	0.082	0.110	
R_2	31.452	31.452	31.452	94.357	1
C_1	C	C	C	0.25C	
R_1	0.225	0.270	0.307	0.307	
R_2	6.640	11.066	24.345	24.345	2
C_1	C	0.5C	0.2C	0.2C	
R_1	0.307	0.369	0.419	0.419	
R_2	1.293	2.156	4.742	4.742	3
C_1	C	0.5C	0.2C	0.2C	

[a] Resistances in kilohms for a K parameter of 1.

Table 3-41. Sixth-Order High-Pass Chebyshev
Cascaded MFB Filter Designs (1 dB)

	Circuit Element Values[a]				
Gain	1	4	25	100	Stage
R_1	0.066	0.066	0.066	0.088	
R_2	38.037	38.037	38.037	114.112	1
C_1	C	C	C	0.25C	
R_1	0.180	0.216	0.246	0.246	
R_2	7.838	13.063	28.738	28.738	2
C_1	C	0.5C	0.2C	0.2C	
R_1	0.246	0.295	0.336	0.336	
R_2	1.283	2.138	4.704	4.704	3
C_1	C	0.5C	0.2C	0.2C	

[a] Resistances in kilohms for a K parameter of 1.

Table 3-42. Sixth-Order High-Pass Chebyshev
Cascaded MFB Filter Designs (2 dB)

Gain	Circuit Element Values[a]				Stage
	1	4	25	100	
R_1	0.050	0.050	0.050	0.066	
R_2	49.093	49.093	49.093	147.278	1
C_1	C	C	C	0.25C	
R_1	0.136	0.163	0.186	0.186	
R_2	9.914	16.523	36.351	36.351	2
C_1	C	0.5C	0.2C	0.2C	
R_1	0.186	0.223	0.254	0.254	
R_2	1.361	2.268	4.990	4.990	3
C_1	C	0.5C	0.2C	0.2C	

[a] Resistances in kilohms for a K parameter of 1.

Table 3-43. Sixth-Order High-Pass Chebyshev
Cascaded MFB Filter Designs (3 dB)

Gain	Circuit Element Values[a]				Stage
	1	4	25	100	
R_1	0.041	0.041	0.041	0.054	
R_2	59.626	59.626	59.626	178.880	1
C_1	C	C	C	0.25C	
R_1	0.111	0.133	0.151	0.151	
R_2	11.927	19.879	43.733	43.733	2
C_1	C	0.5C	0.2C	0.2C	
R_1	0.151	0.182	0.206	0.206	
R_2	1.486	2.477	5.448	6.448	3
C_1	C	0.5C	0.2C	0.2C	

[a] Resistances in kilohms for a K parameter of 1.

Table 3-44. Eighth-Order High-Pass Butterworth
Cascaded MFB Filter Designs

Gain	Circuit Element Values[a]				Stage
	1	4	40	100	
R_1	0.207	0.207	0.248	0.248	
R_2	12.237	12.237	20.395	20.395	1
C_1	C	C	0.5C	0.5C	
R_1	0.589	0.589	0.707	0.707	
R_2	4.297	4.297	7.162	7.162	2
C_1	C	C	0.5C	0.5C	
R_1	0.882	1.059	1.059	1.203	
R_2	2.871	4.785	4.785	10.528	3
C_1	C	0.5C	0.5C	0.2C	
R_1	1.041	1.249	1.419	1.419	
R_2	2.434	4.057	8.925	8.925	4
C_1	C	0.5C	0.2C	0.2C	

[a] Resistances in kilohms for a K parameter of 1.

Table 3-45. Eighth-Order High-Pass Chebyshev Cascaded MFB Filter Designs (0.1 dB)

Gain	Circuit Element Values[a]				Stage
	1	4	40	100	
R_1	0.068	0.068	0.081	0.081	
R_2	39.909	39.909	66.516	66.516	1
C_1	C	C	0.5C	0.5C	
R_1	0.193	0.193	0.232	0.232	
R_2	10.468	10.468	17.447	17.447	2
C_1	C	C	0.5C	0.5C	
R_1	0.289	0.347	0.347	0.395	
R_2	3.644	6.074	6.074	13.362	3
C_1	C	0.5C	0.5C	0.2C	
R_1	0.341	0.410	0.465	0.465	
R_2	1.081	1.801	3.963	3.963	4
C_1	C	0.5C	0.2C	0.2C	

[a] Resistances in kilohms for a K parameter of 1.

Table 3-46.   Eighth-Order High-Pass Chebyshev
Cascaded MFB Filter Designs (0.5 dB)

Gain	\multicolumn{4}{c}{Circuit Element Values[a]}	Stage
	1	4	40	100	
R_1	0.046	0.046	0.056	0.056	
R_2	55.383	55.383	92.305	92.305	1
C_1	C	C	0.5C	0.5C	
R_1	0.132	0.132	0.158	0.158	
R_2	14.247	14.247	23.746	23.746	2
C_1	C	C	0.5C	0.5C	
R_1	0.197	0.237	0.237	0.269	
R_2	4.606	7.676	7.676	16.887	3
C_1	C	0.5C	0.5C	0.2C	
R_1	0.233	0.279	0.317	0.317	
R_2	0.959	1.598	3.515	3.515	4
C_1	C	0.5C	0.2C	0.2C	

[a] Resistances in kilohms for a K parameter of 1.

Table 3-47. Eighth-Order High-Pass Chebyshev
Cascaded MFB Filter Designs (1 dB)

	Circuit Element Values[a]				
Gain	1	4	40	100	Stage
R_1	0.037	0.037	0.045	0.045	
R_2	67.794	67.794	112.990	112.990	1
C_1	C	C	0.5C	0.5C	
R_1	0.106	0.106	0.127	0.127	
R_2	17.326	17.326	28.877	28.877	2
C_1	C	C	0.5C	0.5C	
R_1	0.158	0.190	0.190	0.216	
R_2	5.454	9.090	9.090	19.998	3
C_1	C	0.5C	0.5C	0.2C	
R_1	0.187	0.224	0.255	0.255	
R_2	0.953	1.588	3.495	3.495	4
C_1	C	0.5C	0.2C	0.2C	

[a] Resistances in kilohms for a K parameter of 1.

Table 3-48. Eighth-Order High-Pass Chebyshev
Cascaded MFB Filter Designs (2 dB)

Gain	1	4	40	100	Stage
R_1	0.028	0.028	0.034	0.034	
R_2	88.346	88.346	147.243	147.243	1
C_1	C	C	0.5C	0.5C	
R_1	0.080	0.080	0.096	0.096	
R_2	22.460	22.460	37.434	37.434	2
C_1	C	C	0.5C	0.5C	
R_1	0.120	0.144	0.144	0.163	
R_2	6.916	11.527	11.527	25.359	3
C_1	C	0.5C	0.5C	0.2C	
R_1	0.141	0.170	0.193	0.193	
R_2	1.013	1.688	3.713	3.713	4
C_1	C	0.5C	0.2C	0.2C	

Circuit Element Values[a]

[a] Resistances in kilohms for a K parameter of 1.

Table 3-49.   Eighth-Order High-Pass Chebyshev
Cascaded MFB Filter Designs (3 dB)

Gain	Circuit Element Values[a]				Stage
	1	4	40	100	
R_1	0.023	0.023	0.027	0.027	
R_2	107.779	107.779	179.631	179.631	1
C_1	C	C	0.5C	0.5C	
R_1	0.065	0.065	0.078	0.078	
R_2	27.334	27.334	45.557	45.557	2
C_1	C	C	0.5C	0.5C	
R_1	0.098	0.117	0.117	0.133	
R_2	8.330	13.883	13.883	30.543	3
C_1	C	0.5C	0.5C	0.2C	
R_1	0.115	0.138	0.157	0.157	
R_2	1.107	1.845	4.058	4.058	4
C_1	C	0.5C	0.2C	0.2C	

[a] Resistances in kilohms for a K parameter of 1.

3.9 Summary of Biquad High-Pass Filter Design Procedure

General circuit

Procedure

Given cutoff f_c (hertz), filter gain, order n, and filter type (Butterworth or Chebyshev), perform the following steps for a second-order filter, or for each stage of a higher-order cascaded filter ($n = 4, 6, 8$):

1. Select a value of capacitance C and determine a K-parameter from

$$K = \frac{100}{f_c C'}$$

where C' is the value of C in microfarads. Alternatively, K may be found from Fig. 3-15a, b, or c. For higher-order designs (say, $n > 4$), it is better to use the equation since greater accuracy is required.

2. Find the resistance values from the appropriate one of Tables 3-50 through 3-53. The resistances in the tables are given for $K = 1$ and hence their values must be multiplied by the K parameter of step 1 to yield the resistances of the circuit. The number G in the tables is the stage gain, and for $n > 2$, the filter gain is the product of the stage gains. The stage gains, which are not necessarily equal, are chosen by the designer.

3. Select standard resistance values that are as close as possible to those indicated by the table and construct the filter, or its stages, in accordance with the general circuit.

Comments and Suggestions

The comments and suggestions for the VCVS high-pass filter given in Sec. 3.7 apply as follows:

(a) Paragraphs (a), (d), and (f) are directly applicable.

(b) Paragraphs (c) and (e) do not apply.

(c) Paragraph (b) applies with the exception that R_{eq} is the resistor connected in the inverting input of the op-amps.

In addition, the following applies:

(d) The stage gain is proportional to $1/R_2$.

(e) The filter response is readily tuned by varying R_2, R_3, and R_4. Varying R_2 affects the gain, varying R_3 affects the passband response, and varying R_4 changes f_c.

A discussion of the biquad high-pass filters was given in Sec. 3.4.

Table 3-50.   Second-Order High-Pass Biquad Filter Designs

		Circuit Element Values[a]				
		Chebyshev				
	Butterworth	0.1 dB	0.5 dB	1 dB	2 dB	3 dB
R_1	1.125/G	0.671/G	1.116/G	1.450/G	1.980/G	2.468/G
R_2	1.592/G	1.592/G	1.592/G	1.592/G	1.592/G	1.592/G
R_3	1.125	2.223	1.693	1.598	1.630	1.747
R_4	1.592	5.273	2.413	1.755	1.310	1.127
R_5	1.592	1.592	1.592	1.592	1.592	1.592

[a] Resistances in kilohms for a K parameter of 1, G = gain.

Table 3-51. Fourth-Order High-Pass Cascaded Biquad Filter Designs

				Circuit Element Values[a]			
				Chebyshev			
	Butterworth	0.1 dB	0.5 dB	1 dB	2 dB	3 dB	Stage
R_1	$2.079/G$	$3.013/G$	$4.538/G$	$5.703/G$	$7.587/G$	$9.343/G$	
R_2	$1.592/G$	$1.592/G$	$1.592/G$	$1.592/G$	$1.592/G$	$1.592/G$	
R_3	2.079	4.006	4.826	5.626	7.046	8.438	1
R_4	1.592	2.116	1.693	1.570	1.478	1.437	
R_5	1.592	1.592	1.592	1.592	1.592	1.592	
R_1	$0.861/G$	$1.248/G$	$1.880/G$	$2.362/G$	$3.143/G$	$3.870/G$	
R_2	$1.592/G$	$1.592/G$	$1.592/G$	$1.592/G$	$1.592/G$	$1.592/G$	
R_3	0.861	0.777	0.670	0.660	0.696	0.758	2
R_4	1.592	0.991	0.567	0.445	0.353	0.312	
R_5	1.592	1.592	1.592	1.592	1.592	1.592	

[a] Resistances in kilohms for a K parameter of 1, G = stage gain, filter gain = product of stage gains.

Table 3-52. Sixth-Order High-Pass Cascaded Biquad Filter Designs

			Circuit Element Values[a]				
				Chebyshev			
	Butterworth	0.1 dB	0.5 dB	1 dB	2 dB	3 dB	Stage
R_1	$3.075/G$	$6.938/G$	$10.248/G$	$12.798/G$	$16.941/G$	$20.816/G$	
R_2	$1.592/G$	$1.592/G$	$1.592/G$	$1.592/G$	$1.592/G$	$1.592/G$	
R_3	3.075	7.837	10.484	12.679	16.364	19.875	1
R_4	1.592	1.798	1.628	1.577	1.537	1.520	
R_5	1.592	1.592	1.592	1.592	1.592	1.592	
R_1	$1.125/G$	$2.539/G$	$3.751/G$	$4.684/G$	$6.201/G$	$7.619/G$	
R_2	$1.592/G$	$1.592/G$	$1.592/G$	$1.592/G$	$1.592/G$	$1.592/G$	
R_3	1.125	1.769	2.213	2.613	3.305	3.976	2
R_4	1.592	1.108	0.939	0.888	0.848	0.830	
R_5	1.592	1.592	1.592	1.592	1.592	1.592	
R_1	$0.824/G$	$1.859/G$	$2.746/G$	$3.429/G$	$4.539/G$	$5.578/G$	
R_2	$1.592/G$	$1.592/G$	$1.592/G$	$1.592/G$	$1.592/G$	$1.592/G$	
R_3	0.824	0.490	0.431	0.428	0.454	0.495	3
R_4	1.592	0.419	0.250	0.198	0.159	0.141	
R_5	1.592	1.592	1.592	1.592	1.592	1.592	

[a] Resistances in kilohms for a K parameter of 1, G = stage gain, filter gain = product of stage gains.

Table 3-53. Eighth-Order High-Pass Cascaded Biquad Filter Designs

			Circuit Element Values[a]				
			Chebyshev				
	Butterworth	0.1 dB	0.5 dB	1 dB	2 dB	3 dB	Stage
R_1	4.079/G	12.437/G	18.243/G	22.731/G	30.038/G	36.879/G	
R_2	1.592/G	1.592/G	1.592/G	1.592/G	1.592/G	1.592/G	
R_3	4.079	13.303	18.461	22.598	29.448	35.927	1
R_4	1.592	1.702	1.611	1.582	1.560	1.550	
R_5	1.592	1.592	1.592	1.592	1.592	1.592	
R_1	1.432/G	4.367/G	6.406/G	7.982/G	10.548/G	12.950/G	
R_2	1.592/G	1.592/G	1.592/G	1.592/G	1.592/G	1.592/G	
R_3	1.432	3.489	4.749	5.775	7.487	9.111	2
R_4	1.592	1.272	1.180	1.152	1.130	1.120	
R_5	1.592	1.592	1.592	1.592	1.592	1.592	
R_1	0.957/G	2.918/G	4.280/G	5.333/G	7.048/G	8.653/G	
R_2	1.592/G	1.592/G	1.592/G	1.592/G	1.592/G	1.592/G	
R_3	0.957	1.215	1.535	1.818	2.305	2.777	3
R_4	1.592	0.663	0.571	0.542	0.521	0.511	
R_5	1.592	1.592	1.592	1.592	1.592	1.592	
R_1	0.811/G	2.474/G	3.629/G	4.521/G	5.975/G	7.336/G	
R_2	1.592/G	1.592/G	1.592/G	1.592/G	1.592/G	1.592/G	
R_3	0.811	0.360	0.320	0.318	0.338	0.369	4
R_4	1.592	0.232	0.140	0.112	0.090	0.080	
R_5	1.592	1.592	1.592	1.592	1.592	1.592	

[a] Resistances in kilohms for a K parameter of 1, G = stage gain, filter gain = product of stage gains.

3.10 Summary of Multiple-Feedback High-Pass Filter Design Procedure

The general circuits of the multiple-feedback high-pass filters are given for $n = 3$ through 8 in Figs. 3-8 through 3-13, respectively.

Procedure

Given cutoff f_c (hertz), gain G, order n, and filter type (Butterworth or Chebyshev), perform the following steps.

1. Select a value of capacitance C and determine a K parameter from

$$K = \frac{100}{f_c C'}$$

where C' is the value of C in microfarads. Alternatively, K may be found from Fig. 3-15 a, b, or c. For higher-order designs (say, $n > 4$), it is better to use the equation since greater accuracy is required.

2. Find the remaining element values from the appropriate one of Tables 3-54 through 3-69 as follows. The values of the capacitances other than C are determined directly from the tables using the chosen value of C. The resistances in the tables are given for $K = 1$, and hence their values must be multiplied by the K parameter of step 1 to yield the resistances of the circuit.

3. Select standard resistance values that are as close as possible to those indicated by the table and construct the filter in accordance with the general circuit. In case the remaining capacitances are multiples of C such as 0.47 and so forth, standard values result if C is chosen as a power of 10 (i.e., 0.1, 1, 10, etc.) μF.

Comments and Suggestions

The comments and suggestions for the VCVS high-pass filter given in Sec. 3.7 apply as follows:

(a) Paragraphs (d) and (f) apply directly.

(b) Paragraph (a) does not apply.

(c) Paragraph (b) applies with the exception that R_{eq} for each op-amp is the resistor connected to the noninverting input.

(d) Paragraph (c) applies for each op-amp except that R_3 and R_4 are replaced by the two resistors connected to the inverting input.

(e) Paragraph (e) applies for each VCVS that has gain-setting resistors. In each case R_3 and R_4 are replaced by these gain-setting resistors.

Specific examples of various-order multiple feedback high-pass filters were given in Sec. 3.6.

Table 3-54. Third-Order High-Pass Butterworth Multiple-Feedback Filter Designs

	Circuit Element Values[a]			
Gain	1	2	6	10
R_1	1.143	0.933	0.722	0.635
R_2	0.449	1.836	4.197	5.844
R_3	7.861	2.354	1.331	1.086
R_4	Open	4.708	1.597	1.207
R_5	0	4.708	7.984	10.862

[a] Resistances in kilohms for a K parameter of 1.

Table 3-55. Third-Order High-Pass Chebyshev Multiple-Feedback Filter Designs (0.1 dB)

	Circuit Element Values[a]			
Gain	1	2	6	10
R_1	1.211	1.001	0.802	0.716
R_2	0.332	2.108	4.922	6.861
R_3	16.423	3.131	1.673	1.342
R_4	Open	6.262	2.007	1.491
R_5	0	6.262	10.035	13.415

[a] Resistances in kilohms for a K parameter of 1.

Table 3-56. Third-Order High-Pass Chebyshev Multiple-Feedback Filter Designs (0.5 dB)

	Circuit Element Values[a]			
Gain	1	2	6	10
R_1	0.831	0.694	0.570	0.515
R_2	0.166	1.601	3.780	5.270
R_3	20.880	2.597	1.341	1.064
R_4	Open	5.193	1.609	1.182
R_5	0	5.193	8.044	10.639

[a] Resistances in kilohms for a K parameter of 1.

Table 3-57. Third-Order High-Pass Chebyshev Multiple-Feedback Filter Designs (1 dB)

Gain	Circuit Element Values[a]			
	1	2	6	10
R_1	0.679	0.572	0.476	0.434
R_2	0.108	1.427	3.383	4.714
R_3	27.107	2.426	1.229	0.969
R_4	Open	4.852	1.475	1.077
R_5	0	4.852	7.374	9.691

[a] Resistances in kilohms for a K parameter of 1.

Table 3-58. Third-Order High-Pass Chebyshev Multiple-Feedback Filter Designs (2 dB)

Gain	Circuit Element Values[a]			
	1	2	6	10
R_1	0.528	0.451	0.383	0.352
R_2	0.059	1.278	3.036	4.224
R_3	42.253	2.285	1.133	0.886
R_4	Open	4.570	1.360	0.985
R_5	0	4.570	6.797	8.865

[a] Resistances in kilohms for a K parameter of 1.

Table 3-59. Third-Order High-Pass Chebyshev Multiple-Feedback Filter Designs (3 dB)

Gain	Circuit Element Values[a]			
	1	2	6	10
R_1	0.439	0.379	0.327	0.302
R_2	0.037	1.203	2.855	3.965
R_3	62.823	2.214	1.083	0.843
R_4	Open	4.428	1.299	0.936
R_5	0	4.428	6.496	8.427

[a] Resistances in kilohms for a K parameter of 1.

Table 3-60. Fourth-Order High-Pass Butterworth
Multiple-Feedback Filter Designs

Gain	Circuit Element Values[a]			
	1	2	6	10
R_1	1.663	1.536	1.206	1.011
R_2	1.538	1.557	1.648	1.734
R_3	4.557	1.906	1.093	0.899
R_4	0.604	1.546	3.262	4.513
R_5	15.915	15.915	15.915	15.915
R_6	Open	3.812	1.311	0.999
R_7	0	3.812	6.556	8.992

[a] Resistances in kilohms for a K parameter of 1.

Table 3-61. Fourth-Order High-Pass Chebyshev
Multiple-Feedback Filter Designs (0.1 dB)

Gain	Circuit Element Values[a]			
	1	2	6	10
R_1	1.524	1.431	1.166	0.994
R_2	1.063	1.095	1.204	1.294
R_3	8.948	2.353	1.223	0.974
R_4	0.391	1.541	3.331	4.586
R_5	15.915	15.915	15.915	15.915
R_6	Open	4.706	1.467	1.083
R_7	0	4.706	7.335	9.744

[a] Resistances in kilohms for a K parameter of 1.

Table 3-62. Fourth-Order High-Pass Chebyshev
Multiple-Feedback Filter Designs (0.5 dB)

Gain	Circuit Element Values[a]			
	1	2	6	10
R_1	1.321	1.254	1.049	0.909
R_2	0.698	0.722	0.803	0.870
R_3	10.746	2.147	1.069	0.838
R_4	0.256	1.308	2.836	3.874
R_5	·15.915	15.915	15.915	15.915
R_6	Open	4.294	1.283	0.931
R_7	0	4.294	6.415	8.377

[a] Resistances in kilohms for a K parameter of 1.

Table 3-63. Fourth-Order High-Pass Chebyshev
Multiple-Feedback Filter Designs (1 dB)

Gain	Circuit Element Values[a]			
	1	2	6	10
R_1	1.304	1.239	1.041	1.065
R_2	0.553	0.574	0.646	0.628
R_3	12.580	2.094	1.023	0.759
R_4	0.202	1.230	2.675	3.550
R_5	15.915	15.915	15.915	31.830
R_6	Open	4.189	1.228	0.844
R_7	0	4.189	6.138	7.595

[a] Resistances in kilohms for a K parameter of 1.

Table 3-64.   Fourth-Order High-Pass Chebyshev Multiple-Feedback Filter Designs (2 dB)

Gain	Circuit Element Values[a]			
	1	2	6	10
R_1	1.377	1.348	1.090	1.325
R_2	0.414	0.418	0.497	0.420
R_3	15.956	2.036	0.989	0.710
R_4	0.149	1.167	2.544	3.345
R_5	15.915	31.830	15.915	159.155
R_6	Open	4.072	1.186	0.789
R_7	0	4.072	5.932	7.103

[a] Resistances in kilohms for a K parameter of 1.

Table 3-65.   Fourth-Order High-Pass Chebyshev Multiple-Feedback Filter Designs (3 dB)

Gain	Circuit Element Values[a]			
	1	2	6	10
R_1	1.499	1.506	1.167	1.477
R_2	0.335	0.330	0.413	0.335
R_3	19.405	2.012	0.974	0.700
R_4	0.119	1.138	2.480	3.279
R_5	15.915	159.155	15.915	318.310
R_6	Open	4.025	1.169	0.778
R_7	0	4.025	5.844	7.004

[a] Resistances in kilohms for a K parameter of 1.

Table 3-66. Fifth-Order High-Pass Multiple-Feedback Filter Designs

	Circuit Element Values[a]								
			Chebyshev						
	Butterworth		0.1 dB		0.5 dB		1 dB	2 dB	3 dB
Gain	2	10	2	10	2	10	2	2	2
R_1	5.323	2.670	4.240	2.766	4.349	2.812	5.096	7.284	6.160
R_2	1.816	1.939	0.861	1.111	0.499	0.697	0.347	0.200	0.184
R_3	0.432	0.425	0.394	0.323	0.328	0.272	0.289	0.242	0.227
R_4	2.752	1.018	2.858	1.008	2.561	0.900	2.442	2.327	2.299
R_5	1.395	4.826	1.294	4.470	1.159	3.983	1.113	1.079	1.115
R_6	3.183	15.915	3.183	15.915	3.183	15.915	3.183	3.183	6.366
R_7	5.503	1.132	5.715	1.120	5.122	1.000	4.883	4.654	4.600
R_8	5.503	10.184	5.715	10.080	5.122	9.000	4.883	4.654	4.600

[a] Resistances in kilohms for a K parameter of 1.

Table 3-67. Sixth-Order High-Pass Multiple-Feedback Filter Designs

	Circuit Element Values[a]					
		Chebyshev				
	Butterworth	0.1 dB	0.5 dB	1 dB	2 dB	3 dB
Gain	2	2	2	2	2	2
R_1	4.005	3.426	2.290	2.241	2.123	2.470
R_2	2.976	3.376	3.319	3.399	3.405	3.546
R_3	0.822	0.675	0.707	0.696	0.768	0.734
R_4	1.100	0.543	0.592	0.554	0.529	0.441
R_5	1.608	1.413	1.315	1.290	1.259	1.257
R_6	1.512	0.775	0.516	0.424	0.318	0.290
R_7	4.135	2.622	2.005	1.716	1.615	1.416
R_8	3.183	3.183	5.551	5.064	6.615	3.600
R_9	5.960	6.760	6.640	6.800	6.810	7.100

[a] Resistances in kilohms for a K parameter of 1.

Table 3-68. Seventh-Order High-Pass Multiple-Feedback Filter Designs

Circuit Element Values[a]

| | Butterworth | Chebyshev | | | | |
		0.1 dB	0.5 dB	1 dB	2 dB	3 dB
Gain	2	2	2	2	2	2
R_1	9.137	5.241	2.897	3.080	1.991	3.442
R_2	0.398	0.279	0.407	0.343	0.290	0.308
R_3	2.139	2.882	3.348	3.578	3.859	3.350
R_4	1.294	1.290	1.285	1.285	1.292	1.231
R_5	0.668	0.454	0.331	0.259	0.253	0.178
R_6	2.383	2.236	2.640	2.675	1.988	4.653
R_7	1.815	0.533	0.335	0.321	0.480	0.139
R_8	13.455	5.209	2.087	1.499	1.024	2.028
R_9	4.160	4.906	4.495	2.645	8.725	3.570
R_{10}	4.280	5.760	6.700	7.160	7.720	6.700

[a] Resistances in kilohms for a K parameter of 1.

Table 3-69. Eighth-Order High-Pass Multiple-Feedback Filter Designs

Circuit Element Values[a]

| | Butterworth | Chebyshev | | | | |
		0.1 dB	0.5 dB	1 dB	2 dB	3 dB
Gain	2	2	2	2	2	2
R_1	6.294	3.339	2.204	1.923	1.773	1.677
R_2	0.235	0.305	0.454	0.677	0.537	0.596
R_3	5.715	4.339	3.932	4.298	3.956	4.197
R_4	1.904	1.409	1.307	1.357	1.276	1.298
R_5	0.402	0.408	0.433	0.368	0.461	0.455
R_6	1.090	0.605	0.423	0.389	0.280	0.252
R_7	3.505	3.200	4.426	6.516	6.549	6.371
R_8	2.623	0.685	0.412	0.277	0.262	0.266
R_9	0.760	0.873	0.891	0.600	0.728	0.613
R_{10}	1.879	3.399	3.515	1.290	6.574	7.820
R_{11}	1.608	2.118	3.602	6.164	2.703	2.228
R_{12}	11.430	8.680	7.860	8.600	7.910	8.390

[a] Resistances in kilohms for a K parameter of 1.

4

Bandpass Filters

4.1 General Theory

A bandpass filter passes a band of frequencies of bandwidth B centered approximately about a *center frequency* $\omega_0 \neq 0$, and attenuates all other frequencies. Both B and ω_0 may be measured in radians/second or B may be given in hertz with a center frequency $f_0 = \omega_0/2\pi$ Hz. The ratio, $Q = \omega_0/B$, or if B is in hertz, $Q = f_0/B$, is the *quality factor* and is a measure of the selectivity of the filter. A high Q, for example, indicates a very selective filter since, in this case, the band of frequencies that passes is narrow compared to the center frequency. The *gain* of the filter is defined as the amplitude of its transfer function at the center frequency.

An ideal bandpass response is shown in Fig. 4-1, represented by the broken line. A realizable approximation to the ideal is shown as a solid line. There are two cutoff points, ω_1 and ω_2, as shown and, as in the low- and high-pass cases, cutoff is defined to occur at points where $|H(j\omega)|$ reaches $1/\sqrt{2}$ times its maximum value. For the realizable approximation the bandwidth is then $B = \omega_1 - \omega_2$.

As for the high-pass filter, the bandpass filter transfer function may be obtained by transforming that of a low-pass prototype. The bandpass transfer function obtained in this way is given in the general case by [22]

$$\frac{V_2}{V_1} = \frac{Gb_0}{S^n + b_{n-1}S^{n-1} + \cdots + b_1S + b_0}\bigg|_{S = (s^2 + \omega^2_0)/Bs} \tag{4.1}$$

where, before the indicated substitution is made, the transfer function is

134

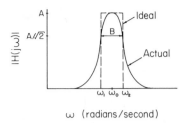

Figure 4-1.   Bandpass amplitude responses.

that of the nth order low-pass filter of Eq. (2.2). Evidently a $2n$th order bandpass function arises from an nth order low-pass function, with a gain of G, as in the low-pass case ($s = j\omega_0$ corresponds to $S = 0$). Also, the filter is a Butterworth or Chebyshev bandpass filter when its low-pass prototype is Butterworth or Chebyshev.

A Butterworth bandpass filter has monotonic amplitude response with a maximally flat passband. In the case of Chebyshev bandpass filters, the passband has ripples, as in the low-pass case, and the rate of attenuation at the cutoff points is much greater than that of the Butterworth. Again, except in the 3 dB ripple case, the frequencies ω_1 and ω_2 are the terminal frequencies of the passband ripple channel in the Chebyshev filter and are the conventional cutoff points in the Butterworth filter. In general, for Eq. (4.1), the center frequency is the geometric mean of the frequencies ω_1 and ω_2. That is, $\omega_0 = \sqrt{\omega_1\omega_2}$, or in hertz, $f_0 = \sqrt{f_1 f_2}$.

Since first-order Butterworth and Chebyshev low-pass amplitude responses have the same form except for scaling, a second-order bandpass function in either case is given from Eq. (4.1) by taking $b_0 = 1$ and $n = 1$. The result is

$$\frac{V_2}{V_1} = \frac{GBs}{s^2 + Bs + \omega_0{}^2} \qquad (4.2)$$

Higher-order responses, for which the Butterworth and Chebyshev cases are different, are given by Eq. (4.1) for $n = 2, 3, \ldots$.

Amplitude and phase responses of a Chebyshev and Butterworth filter of fourth order are shown for comparative purposes in Figs. 4-2 and 4-3.

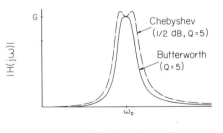

Figure 4-2.   Amplitude   responses   of fourth-order bandpass filters.

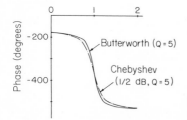

Figure 4-3. Phase responses of fourth-order bandpass filters.

In this chapter, we consider a number of bandpass filter realizations and present rapid design techniques for their construction. The procedure is identical to that of the low-pass and high-pass filters and will be summarized for each filter type at the end of the chapter.

4.2 Cascading of Second-Order Sections

From Eq. (4.1), it may be seen that a $2n$th order bandpass filter transfer function is a ratio of a constant times s^n to a $2n$th degree polynomial in s, and thus is factorable into n factors similar to the function of Eq. (4.2). Therefore, higher-order filters may be obtained by cascading second-order sections. We shall consider this method later in the design procedure of the biquad bandpass filter to be discussed in Sec. 4.6.

We note that a fourth-order bandpass transfer function has the form of the product of two second-order bandpass transfer functions, but it also can be factored into a product of a low-pass transfer function and a high-pass transfer function. Thus, a quick way to get a variety of fourth-order bandpass filters (although not of the Butterworth or Chebyshev type) is to cascade a low-pass filter with cutoff point f_2 with a high-pass filter of cutoff point $f_1 < f_2$, making use of the results of Chapters 2 and 3. The result is a filter with center frequency approximately $f_0 = \sqrt{f_1 f_2}$ (it is exactly this if both filters in cascade are Butterworth types [1]) and B approximately $f_2 - f_1$. The approximations improve as the difference $f_2 - f_1$ increases, and hence as Q decreases. Best results are obtained if f_1 and f_2 are at least one octave apart [34], and hence for Q not exceeding approximately $\sqrt{2}$. (An octave is the interval between two frequencies, one twice the other.) Sharper cutoff features may be obtained by cascading fourth-order low- and high-pass filters, of course, but again the results are not of the Butterworth or Chebyshev eighth-order types.

In the case where both the low- and high-pass filters cascaded are Butterworth filters, if $f_1 = f_2$, then the common value is also f_0 and $Q = 1.1$. If $f_2 = 2f_1$ (one octave apart), then $f_0 = \sqrt{2}f_1$ and $Q = 0.912$ [1].

A much sharper bandpass filter may be obtained by cascading two identical bandpass filters. Again the result, however, is not a fourth-order Butterworth or Chebyshev type. If Q_1 is the quality factor of each stage and there are n stages, then Q of the filter is $Q_1/\sqrt{\sqrt[n]{2} - 1}$ [1]. These values and the corresponding bandwidths are shown in Table 4-1, for $n = 1, 2, 3, 4, 5$, and B_1 the bandwidth of a single stage.

Table 4-1. Q Values for n Cascaded Identical Bandpass Filters

n	Bandwidth	Q
1	B_1	Q_1
2	$0.644B_1$	$1.55Q_1$
3	$0.510B_1$	$1.96Q_1$
4	$0.435B_1$	$2.30Q_1$
5	$0.386B_1$	$2.60Q_1$

4.3   VCVS Bandpass Filters

A circuit that realizes Eq. (4.2), and one of which we shall use to obtain second-order bandpass filters, is the VCVS circuit of Fig. 4-4, credited to Kerwin and Huelsman [35]. Analysis of the circuit shows that Eq. (4.2) is realized if

$$B = \frac{1}{C}\left[\frac{1}{R_1} + \frac{2}{R_2} + \frac{1 - \mu}{R_3}\right]$$

$$\omega_0^2 = \frac{1}{R_2C^2}\left(\frac{1}{R_1} + \frac{1}{R_3}\right) \qquad (4.3)$$

$$G = \frac{\mu}{R_1CB}$$

where

$$\mu = 1 + \frac{R_5}{R_4} \qquad (4.4)$$

The VCVS bandpass filter performs best for values of $Q \leq 10$ [2]. Its advantages are those of a VCVS type, discussed earlier in Sec. 2.5 in the low-pass case. A unique feature, as seen from Eqs. (4.3) is that the bandwidth B can be varied by changing μ without affecting the center frequency ω_0. Tuning can be accomplished by adjusting R_2 for the desired ω_0 and then by adjusting μ for the desired B.

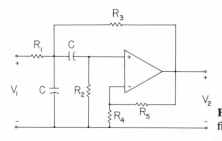

Figure 4-4. A second-order VCVS bandpass filter.

The procedure for constructing a VCVS bandpass filter is very similar to that of the low-pass filter described in Sec. 2.5 of Chapter 2. The procedure is summarized in Sec. 4.9, where the general circuit is shown followed by the design tables. For the VCVS bandpass filter, the pertinent tables are Tables 4-2 through 4-9. The $\dot{K}$ parameter for use with the tables may be calculated from

$$K = \frac{100}{f_0 C'} \tag{4.5}$$

where C' is the value of C in microfarads. Alternatively, K may be read from the appropriate one of Figs. 4-14a, b, or c.

As an example, suppose we want a VCVS bandpass filter with a center frequency $f_0 = 1000$ Hz, a gain $G = 2$, and $Q = 10$. Selecting $C = 0.01$ μF

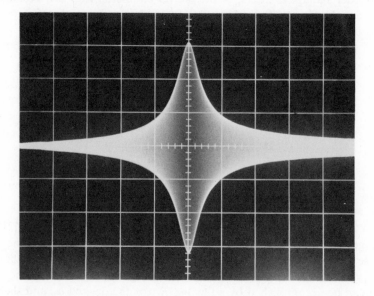

Figure 4-5. A second-order VCVS bandpass amplitude response.

we have from Eq. (4.5) a K parameter of 10. The resistances are given in Table 4-9 for $K = 1$, and thus for our case we must multiply these values by 10. The results in kilohms are $R_1 = 159.15$, $R_2 = 23.32$, $R_3 = 11.66$, $R_4 = R_5 = 46.64$. Standard values of 160, 22, 12, and 47 kΩ were used in the circuit of Fig. 4-4, with the results $f_1 = 974$ Hz, $f_2 = 1074$ Hz (and, hence, $f_0 = 1023$ Hz and $Q = 10.23$), and $G = 2$. The response is shown in Fig. 4-5.

4.4   Infinite-Gain Multiple Feedback Bandpass Filters

Another circuit that realizes the second-order bandpass filter function is the *infinite-gain multiple feedback* (infinite-gain MFB) filter [36] of Fig. 4-6. This filter is analogous to the infinite-gain MFB low-pass filter of Sec. 2.6, and similarly has an inverting gain. Analysis shows that Eq. (4.2) is satisfied if

$$G = -\frac{R_3}{2R_1}$$

$$B = \frac{2}{R_3 C} \tag{4.6}$$

$$\omega_0{}^2 = \frac{1}{R_3 C^2}\left(\frac{1}{R_1} + \frac{1}{R_2}\right)$$

The infinite-gain MFB filter has the advantage of fewer elements than that of the VCVS bandpass filter, as well as other advantages cited for its low-pass counterpart in Sec. 2.6. For high Q, the network of Fig. 4-6 has a wide spread of element values and large Q sensitivities [2]. For this reason it should probably be restricted to values of $Q \leq 10$.

Practical values of the network elements may be obtained using the procedure described in the summary in Sec. 4.10. The pertinent tables are Tables 4-10 through 4-17.

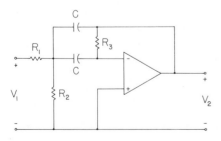

Figure 4-6.   A second-order infinite-gain MFB bandpass filter.

4.5   Positive-Feedback Bandpass Filters

The second-order bandpass filters of the previous two sections are limited, for best results, to Q's of 10 or so with moderate gains. A circuit using two op-amps, which may be used to obtain values of Q up to 50, is the *positive-feedback* (PFB) circuit [2] of Fig. 4-7. (Positive feedback means that the signal fed back, in this case through R_3, is a noninverted signal at ω_0.)

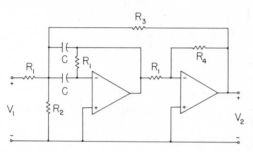

Figure 4-7. A second-order positive-feedback bandpass filter.

Analysis of Fig. 4-7 shows that Eq. (4.2) is satisfied, for C and ω_0 both normalized to 1, with

$$G = \frac{QR_4}{R_1{}^2}$$

$$B = \frac{1}{Q} = \frac{2 - \dfrac{R_4}{R_3}}{R_1} \tag{4.7}$$

$$\omega_0{}^2 = 1 = \frac{\dfrac{1}{R_1} + \dfrac{1}{R_2} + \dfrac{1}{R_3}}{R_1}$$

The denormalization to obtain the desired center frequency and practical element values is identical to that described earlier in Sec. 2.4, except that f_c is replaced by f_0.

The quality factor Q, and hence the bandwidth, can be varied to some degree without appreciably changing f_0 by varying R_4. This is evident from Eq. (4.7) since ω_0 is independent of R_4. In an example given in [1], by varying R_4 by means of a potentiometer, Q was increased from 43.3 to 102. The gain was increased from 4.4 to 12.5 while f_0 changed only slightly, from 2037 Hz to 2035 Hz.

Practical positive-feedback bandpass filter designs may be obtained using Tables 4-18 through 4-27. The general procedure is described in Sec. 4.11.

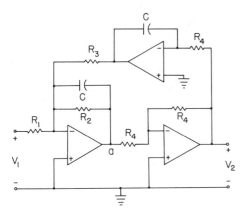

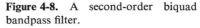

Figure 4-8. A second-order biquad bandpass filter.

4.6   Biquad Bandpass Filters

A biquad circuit, very similar to that of Sec. 2.7 for the low-pass case, which realizes the second-order bandpass function is shown in Fig. 4-8. The biquad version we are considering is that discussed by Brandt [37] and is very similar to that of Tow [29]. Analysis shows that it achieves Eq. (4.2) with

$$G = \frac{R_2}{R_1}$$

$$B = \frac{1}{R_2 C} \tag{4.8}$$

$$\omega_0{}^2 = \frac{1}{R_3 R_4 C^2}$$

If an inverting gain is desired, the output V_2 may be taken at point a, in which case Eqs. (4.8) still hold except that the sign of G is changed.

Although the biquad circuit requires more elements than the other circuits of this chapter, it is a very popular design because of its excellent tuning features. Also, the biquad bandpass filter is capable of attaining Q's of 100 or more, which is, in general, beyond the capability of the other circuits we consider. In addition, the biquad circuit has good stability and cascading of several sections is more feasible for obtaining higher-order Butterworth and Chebyshev responses.

In the case of a second-order biquad bandpass filter the tuning procedure is quite simple. The gain is adjusted by varying R_1, Q is changed by varying R_2, and changing R_3 affects the center frequency.

The design procedure for the biquad bandpass filter is given in Sec. 4.12. Second-order designs can be performed using Table 4-28 and fourth-order designs (cascading of two biquads) are given in Tables 4-29 through 4-34.

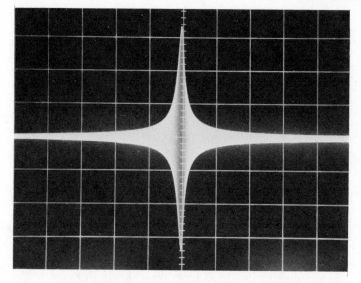

Figure 4-9. A second-order biquad bandpass amplitude response.

As an example, suppose we want a second-order biquad bandpass filter with $f_0 = 1000$ Hz, $Q = 100$, and $G = 10$. Using capacitances of $C = 0.1\ \mu$F we have from Eq. (4.5) a K parameter of 1. From Table 4-28 the other circuit element values are $R_1 = 15.92$, $R_2 = 159.2$, and $R_3 = R_4 = 1.592$ (all kilohms). The actual values used were 15.9, 160, and 1.6 kΩ respectively, yielding the response shown in Fig. 4-9. The results were $f_0 = 1000$ Hz, $Q = 92$, and $G = 9.4$.

4.7 Multiple-Feedback Bandpass Filters

A circuit of our own design [38] which realizes (4.1) for $n = 2$ is the fourth-order *multiple-feedback* bandpass filter of Fig. 4-10. Analysis shows that for ω_0 and C both normalized to 1, the fourth-order realization is achieved with

$$\frac{A}{C_1} + \frac{D}{C_2} = \frac{b_1}{Q}$$

$$\frac{C_1 E + AD + C_2 F + \dfrac{H\mu_2 R_1}{R_7}}{C_1 C_2} = 2 + \frac{b_0}{Q^2}$$

$$\frac{AE + FD}{C_1 C_2} = \frac{b_1}{Q} \qquad (4.9)$$

$$\frac{EF}{C_1 C_2} = 1$$

$$\frac{H}{C_1 C_2} = \frac{G b_0}{Q^2}$$

where

$$\mu_1 = 1 + \frac{R_{10}}{R_9}$$

$$\mu_2 = -\frac{R_{12}}{R_{11}}$$

$$A = C_1\left[\frac{1}{R_1} + \frac{1}{R_2} + \frac{1}{R_7} + \frac{1}{R_3}(1 - \mu_1)\right] + \frac{1}{R_2}$$

$$D = \frac{1}{R_6} + C_2\left(\frac{1}{R_6} + \frac{\mu_2}{R_8}\right)$$ (4.10)

$$E = \frac{1}{R_6}\left(\frac{1}{R_4} + \frac{1}{R_5} + \frac{1}{R_6}\right)$$

$$F = \frac{1}{R_2}\left(\frac{1}{R_1} + \frac{1}{R_3} + \frac{1}{R_7}\right)$$

$$H = \frac{\mu_1 C_1 C_2}{R_1 R_4}$$

As in the low- and high-pass cases, the multiple-feedback bandpass filter presents the mathematical difficulty of solving the set of equations (4.9) and (4.10) for the various cases. The advantages are those of a multiple-feedback structure as cited in Sec. 2.8. Also, the structure of Fig. 4-10 resembles that of the biquad filter of Fig. 4-8, although it has a few more elements. However, the multiple-feedback structure is a fourth-order filter whereas the biquad is only second-order.

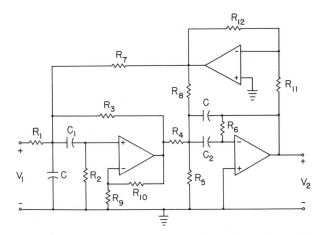

Figure 4-10.   A fourth-order multiple-feedback bandpass filter.

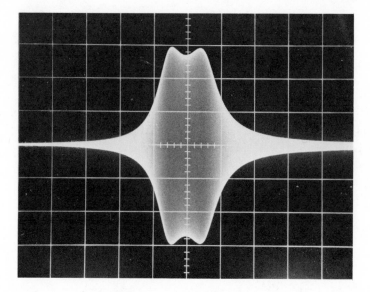

Figure 4-11.   A fourth-order, 0.5 dB, Chebyshev bandpass amplitude response.

The design procedure for the multiple-feedback bandpass filter is given in Sec. 4.13. The pertinent tables are Tables 4-35 through 4.58.

As an example, suppose we want a fourth-order, $\frac{1}{2}$ dB, Chebyshev bandpass filter with $f_0 = 1000$ Hz, $Q = 5$, and $G = 2$. Using $C = 0.01\ \mu$F, we have from Eq. (4.5) a K parameter of 10. The other circuit element values, from Table 4-51, are given by $R_1 = 58.65$, $R_2 = 53.83$, $R_3 = 18.05$, $R_4 = 75.35$, $R_5 = 2.62$, $R_6 = 22.77$, $R_7 = 68.17$, $R_8 = 118.67$, $R_9 = 102.06$, $R_{10} = 113.9$, $R_{11} = 100$, $R_{12} = 758.7$ kΩ, $C_1 = C = 0.01\ \mu$F, and $C_2 = 2C = 0.02\ \mu$F. (The resistances in the table are all multiplied by the K parameter of 10.) Using element values as close as possible to the calculated values, the circuit was constructed with the results $f_0 = 971$ Hz, $Q = 5.2$, $G = 2.2$, with 0.7 dB ripple. The response is shown in Fig. 4-11.

4.8   Multiple-Resonator Bandpass Filters

A multiple-feedback bandpass circuit [39], developed along the lines suggested by Hurtig [40], which realizes filters of orders 2, 4, 6, ..., is shown in Fig. 4-12. The networks labeled N are identical second-order bandpass *resonators* (sections), and to distinguish this filter from those of the previous section, we refer to it as a *multiple-resonator bandpass filter*.

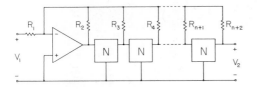

Figure 4-12. A multiple-resonator bandpass filter.

If the number of resonators is n, then the circuit of Fig. 4-12 yields a $2n$th order bandpass filter with transfer function

$$\frac{V_2}{V_1} = \frac{-a_0 H^n}{1 + a_1 H + a_2 H^2 + \cdots + a_n H^n} \tag{4.11}$$

where

$$a_0 = \frac{R_2}{R_1} \tag{4.12}$$

$$a_i = \frac{R_2}{R_{i+2}}; \qquad i = 1, 2, \ldots, n$$

and $H = V_{\text{out}}/V_{\text{in}}$ is the transfer function of resonator N.

For the normalized case ($\omega_0 = 1$), specifying the order and type of bandpass filter determines the coefficients $b_0, b_1, \ldots, b_{n-1}$ ($b_n = 1$) of Eq. (4.1). For a given gain G and quality factor Q, Eqs. (4.11) and (4.12) are satisfied by [39]

$$a_0 = b_0 G$$

$$a_i = \sum_{k=0}^{i} \binom{n-k}{i-k}(-K_1)^{k-i} b_{n-k}; \qquad i = 1, 2, \ldots, n \tag{4.13}$$

where K_1 is the gain of each resonator, given by

$$K_1 = \frac{2n}{b_{n-1}} \tag{4.14}$$

and the resonator quality factor is given by

$$Q_1 = \frac{2nQ}{b_{n-1}} \tag{4.15}$$

The resonators are also normalized to a center frequency $\omega_0 = 1$, and denormalization to the required center frequency can be carried out as described in Sec. 2.4.

The center frequency and Q of the filter are determined by the resonators, and the filter type (Butterworth or Chebyshev) and gain are determined by the feedback network of resistors. The filter has the advantages of multiple

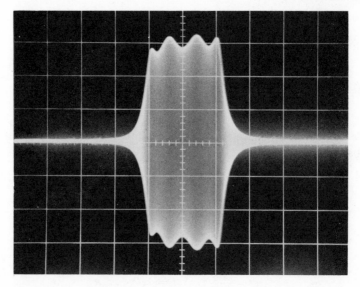

Figure 4-13. An eighth-order, 1 dB, Chebyshev bandpass amplitude response.

feedback, as discussed earlier in Sec. 2.8. It also has the advantage of the cascaded network in that each resonator, or section, can be tuned separately to the desired center frequency f_0 of the filter, the gain K_1, and quality factor Q_1 required by Eqs. (4.14) and (4.15) of the resonators. In addition, the filter gain G can be set by adjusting R_1, as is seen from the first of Eqs. (4.13) and (4.12).

Rapid design techniques for the multiple-resonator bandpass filter are presented in Sec. 4.14, and the pertinent tables are Tables 4-59 through 4.61. For each type and order (fourth, sixth, and eighth), the feedback resistances, the required resonator gain K_1, and quality factor Q_1 are given in the tables. Since R_2 is arbitrary, the resistances given may be multiplied by any convenient scale factor. Each resonator must be designed for the given K_1 and Q_1 and desired center frequency f_0. This may be done quite readily using the biquad circuit procedure of Sec. 4.12. If the given Q, type, and order results in a value Q_1, from Eq. (4.15), which can be attained by a simpler resonator, then the designer can choose his own from some of those presented earlier in the chapter. However, the choice must yield a noninverting gain, as in the case of the VCVS, the biquad, or the positive-feedback bandpass (PFB) filters, for example. For the user's convenience, PFB resonator designs are included in Tables 4-62 through 4-79. These may be used for resonator Q's of approximately 60 or less.

As an example, suppose we want an eighth-order, 1 dB, Chebyshev bandpass filter with $Q = 5$, $G = 1$, and $f_0 = 1000$ Hz. The circuit will be that of Fig. 4-12 with $n = 4$. By Table 4-61 we have $R_1 = 36.281$, $R_2 = 10$, $R_3 = 20.991$, $R_4 = 8.343$, $R_5 = 23.251$, and $R_6 = 48.451$ kΩ. In addition, each resonator must have a gain $K_1 = 8.396$ and $Q_1 = 8.396Q = 41.98$. Thus the PFB resonators of Fig. 4-7 are sufficient. The pertinent design table is Table 4-77, from which the resonator resistances are (for a K parameter of 1) $R_1 = 5.033$, $R_2 = 0.845$, $R_3 = 1.654$, and $R_4 = 3.183$ kΩ. Using $C = 0.01$ μF we have from Eq. (4.5) a K parameter of 10. Thus the resonator resistances must be multiplied by 10. The feedback resistances do not involve the K parameter and may be left as they are or all multiplied by a convenient scale factor. The circuit was constructed using measured 1% resistance values as close as possible to those calculated, with the results $f_0 = 1005$ Hz, $G = 1.1$, $Q = 5.1$, and a ripple of 1 dB measured about f_0. The response is shown in Fig. 4-13.

4.9 Summary of VCVS Bandpass Filter Design Procedure $(Q \leq 10)$

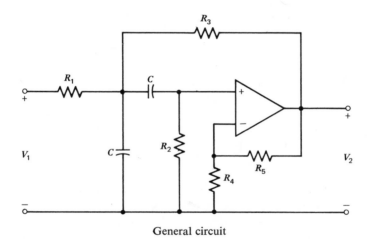

General circuit

Procedure

Given center frequency f_0 (hertz), gain G, and Q (or bandwidth $B = f_0/Q$), perform the following steps for a second-order filter, or for each identical stage of a higher-order cascaded filter.

1. Select a value of capacitance C and determine a K parameter from

$$K = \frac{100}{f_0 C'}$$

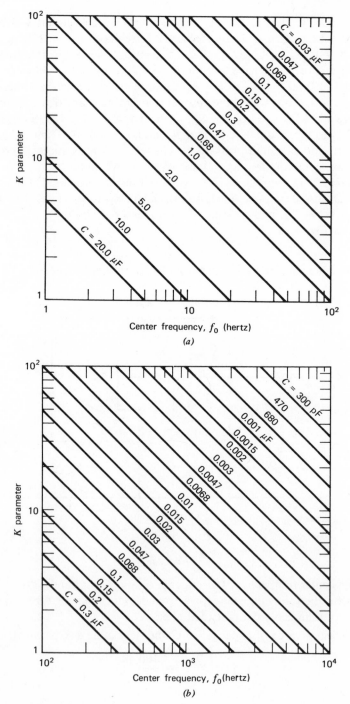

Center frequency, f_0 (hertz)

(a)

Center frequency, f_0(hertz)

(b)

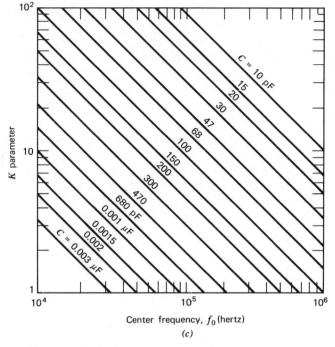

Figure 4-14. (a, b, and c) K parameter versus frequency.

where C' is the value of C in microfarads. Alternatively, K may be found from Fig. 4-14a, b, or c.

2. Find the resistance values from the appropriate one of Tables 4-2 through 4-9. The resistances in the tables are given for $K = 1$, and hence their values must be multiplied by the K parameter of step 1 to yield the resistances of the circuit.

3. Select standard resistance values that are as close as possible to those indicated by the table and construct the filter, or its stages, in accordance with the general circuit.

Comments and Suggestions

(a) The values in the tables for R_4 and R_5 were determined to minimize the dc offset of the op-amp. Other values of R_4 and R_5 may be used as long as the ratio R_5/R_4 is the same as that of the table values.

(b) Standard resistance values of 5% tolerance normally yield acceptable results. In all cases, for best performance, resistance values close to those indicated by the tables should be used.

In the case of capacitors, 5% tolerances should be used for best results.

Since precision capacitors are relatively expensive, it may be desirable to use capacitors of higher tolerances, in which case trimming is generally required. In most cases, 10% capacitors are quite often satisfactory.

(c) The gain of the filter can be adjusted to the correct value by using a potentiometer in lieu of resistors R_4 and R_5. This is accomplished by connecting the center tap of the potentiometer to the inverting input of the op-amp. The center frequency f_0 can be fixed and the bandwidth (or Q) changed by this adjustment.

(d) The open-loop gain of the op-amp should be at least 50 times the gain of the filter at f_a, and the desired peak-to-peak voltage at f_a should not exceed $10^6/\pi f_a$ times the slew rate of the op-amp, where f_a is the highest frequency desired in the passband. Thus, for high values of f_a, externally compensated op-amps may be required.

(e) The values of Q and bandwidth for n identical cascaded sections, $n = 1, 2, 3, 4, 5$, are shown in Table 4-1.

A specific example of a second-order VCVS design was given in Sec. 4.3.

Table 4-2. Second-Order VCVS Bandpass Filter Designs ($Q = 1$)

Gain	Circuit Element Values[a]					
	1	2	4	6	8	10
R_1	3.183	1.592	0.796	0.531	0.398	0.318
R_2	2.251	3.183	5.668	8.550	11.578	14.669
R_3	1.741	1.592	1.019	0.671	0.486	0.377
R_4, R_5	4.502	6.366	11.336	17.110	23.156	29.338

[a] Resistances in kilohms for a K parameter of 1.

Table 4-3. Second-Order VCVS Bandpass Filter Designs ($Q = 2$)

Gain	Circuit Element Values[a]					
	1	2	4	6	8	10
R_1	6.366	3.183	1.592	1.061	0.796	0.637
R_2	2.251	2.684	3.741	4.993	6.366	7.811
R_3	1.367	1.342	1.178	0.972	0.796	0.661
R_4, R_5	4.501	5.368	7.482	9.986	12.732	15.622

[a] Resistances in kilohms for a K parameter of 1.

Table 4-4. Second-Order VCVS Bandpass Filter Designs ($Q = 3$)

	Circuit Element Values[a]					
Gain	1	2	4	6	8	10
R_1	9.549	4.775	2.387	1.592	1.194	0.955
R_2	2.251	2.532	3.183	3.939	4.775	5.668
R_3	1.276	1.266	1.194	1.079	0.955	0.840
R_4, R_5	4.502	5.064	6.366	7.878	9.550	11.336

[a] Resistances in kilohms for a K parameter of 1.

Table 4-5. Second-Order VCVS Bandpass Filter Designs ($Q = 4$)

	Circuit Element Values[a]					
Gain	1	2	4	6	8	10
R_1	12.732	6.366	3.183	2.122	1.592	1.273
R_2	2.251	2.459	2.925	3.456	4.039	4.667
R_3	1.235	1.229	1.189	1.120	1.035	0.946
R_4, R_5	4.502	4.918	5.850	6.912	8.078	9.334

[a] Resistances in kilohms for a K parameter of 1.

Table 4-6. Second-Order VCVS Bandpass Filter Designs ($Q = 5$)

	Circuit Element Values[a]					
Gain	1	2	4	6	8	10
R_1	15.915	7.958	3.979	2.653	1.989	1.592
R_2	2.251	2.416	2.778	3.183	3.626	4.100
R_3	1.211	1.208	1.183	1.137	1.077	1.010
R_4, R_5	4.502	4.832	5.556	6.366	7.252	8.200

[a] Resistances in kilohms for a K parameter of 1.

Table 4-7. Second-Order VCVS Bandpass Filter Designs ($Q = 6$)

Gain	Circuit Element Values[a]					
	1	2	4	6	8	10
R_1	19.099	9.549	4.775	3.183	2.387	1.910
R_2	2.251	2.387	2.684	3.010	3.363	3.741
R_3	1.196	1.194	1.176	1.144	1.100	1.049
R_4, R_5	4.502	4.774	5.368	6.020	6.726	7.482

[a] Resistances in kilohms for a K parameter of 1.

Table 4-8. Second-Order VCVS Bandpass Filter Designs ($Q = 8$)

Gain	Circuit Element Values[a]					
	1	2	4	6	8	10
R_1	25.465	12.732	6.366	4.244	3.183	2.546
R_2	2.251	2.352	2.569	2.802	3.052	3.318
R_3	1.177	1.176	1.167	1.148	1.123	1.090
R_4, R_5	4.502	4.704	5.138	5.604	6.104	6.636

[a] Resistances in kilohms for a K parameter of 1.

Table 4-9. Second-Order VCVS Bandpass Filter Designs ($Q = 10$)

Gain	Circuit Element Values[a]					
	1	2	4	6	8	10
R_1	31.831	15.915	7.958	5.305	3.979	3.183
R_2	2.251	2.332	2.502	2.684	2.876	3.078
R_3	1.167	1.166	1.160	1.148	1.131	1.110
R_4, R_5	4.502	4.664	5.004	5.368	5.752	6.156

[a] Resistances in kilohms for a K parameter of 1.

4.10   Summary of Infinite-Gain MFB Bandpass Filter Design Procedure $(Q \leq 10)$

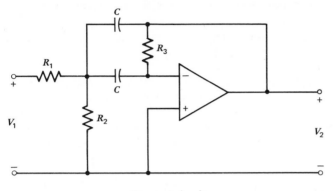

General circuit

Procedure

Given center frequency f_0 (hertz), gain G, and Q (or bandwidth $B = f_0/Q$), perform the following steps for a second-order filter, or for each identical stage of a higher-order cascaded filter.

1.   Select a value of capacitance C and determine a K parameter from

$$K = \frac{100}{f_0 C'}$$

where C' is the value of C in microfarads. Alternately, K may be found from Fig. 4-14a, b, or c.

2.   Find the resistance values from the appropriate one of Tables 4-10 through 4-17. The resistances in the tables are given for $K = 1$ and hence their values must be multiplied by the K parameter of step 1 to yield the resistances of the circuit.

3.   Select standard resistance values that are as close as possible to those indicated by the table and construct the filter, or its stages, in accordance with the general circuit.

Comments and Suggestions

The comments and suggestions for the VCVS bandpass filter given in Sec. 4.9 apply as follows:

(a) Paragraphs (b), (d), and (e) are directly applicable.
(b) Paragraphs (a) and (c) do not apply.

In addition, the following applies:

(c) The inverting gain of the filter is $R_3/2R_1$. Gain adjustments can be made by using a potentiometer in lieu of R_1. This affects f_0, however. Varying R_3 affects Q (or B).

(d) For minimum dc offset, a resistance equal to R_3 can be placed in the noninverting input to ground.

Table 4-10. Second-Order Multiple Feedback Bandpass
Filter Designs ($Q = 2$)

Gain	Circuit Element Values[a]					
	1	2	3	4	6	8
R_1	3.183	1.592	1.061	0.796	0.531	0.398
R_2	0.455	0.531	0.637	0.796	1.592	Open
R_3	6.366	6.366	6.366	6.366	6.366	6.366

[a] Resistances in kilohms for a K parameter of 1.

Table 4-11. Second-Order Multiple-Feedback Bandpass
Filter Designs ($Q = 3$)

Gain	Circuit Element Values[a]					
	1	2	4	6	8	10
R_1	4.775	2.387	1.194	0.796	0.597	0.477
R_2	0.281	0.298	0.341	0.398	0.477	0.597
R_3	9.549	9.549	9.549	9.549	9.549	9.549

[a] Resistances in kilohms for a K parameter of 1.

Table 4-12. Second-Order Multiple-Feedback Bandpass Filter
Designs ($Q = 4$)

Gain	Circuit Element Values[a]					
	1	2	4	6	8	10
R_1	6.366	3.183	1.592	1.061	0.796	0.637
R_2	0.205	0.212	0.227	0.245	0.265	0.289
R_3	12.732	12.732	12.732	12.732	12.732	12.732

[a] Resistances in kilohms for a K parameter of 1.

Table 4-13. Second-Order Multiple-Feedback Bandpass Filter Designs ($Q = 5$)

Gain	Circuit Element Values[a]					
	1	2	4	6	8	10
R_1	7.958	3.979	1.989	1.326	0.995	0.796
R_2	0.162	0.166	0.173	0.181	0.189	0.199
R_3	15.915	15.915	15.915	15.915	15.915	15.915

[a] Resistances in kilohms for a K parameter of 1.

Table 4-14. Second-Order Multiple-Feedback Bandpass Filter Designs ($Q = 6$)

Gain	Circuit Element Values[a]					
	1	2	4	6	8	10
R_1	9.549	4.775	2.387	1.592	1.194	0.955
R_2	0.134	0.136	0.140	0.145	0.149	0.154
R_3	19.099	19.099	19.099	19.099	19.099	19.099

[a] Resistances in kilohms for a K parameter of 1.

Table 4-15. Second-Order Multiple-Feedback Bandpass Filter Designs ($Q = 7$)

Gain	Circuit Element Values[a]					
	1	2	4	6	8	10
R_1	11.141	5.570	2.785	1.857	1.393	1.114
R_2	0.115	0.116	0.119	0.121	0.124	0.127
R_3	22.282	22.282	22.282	22.282	22.282	22.282

[a] Resistances in kilohms for a K parameter of 1.

Table 4-16. Second-Order Multiple-Feedback Bandpass Filter
Designs ($Q = 8$)

	Circuit Element Values[a]					
Gain	1	2	4	6	8	10
R_1	12.732	6.336	3.183	2.122	1.592	1.273
R_2	0.100	0.101	0.103	0.104	0.106	0.108
R_3	25.465	25.465	25.465	25.465	25.465	25.465

[a] Resistances in kilohms for a K parameter of 1.

Table 4-17. Second-Order Multiple-Feedback Bandpass Filter
Designs ($Q = 10$)

	Circuit Element Values[a]					
Gain	1	2	4	6	8	10
R_1	15.915	7.958	3.979	2.653	1.989	1.592
R_2	0.080	0.080	0.081	0.082	0.083	0.084
R_3	31.831	31.831	31.831	31.831	31.831	31.831

[a] Resistances in kilohms for a K parameter of 1.

4.11 Summary of PFB Bandpass Filter Design Procedure ($Q \leq 50$)

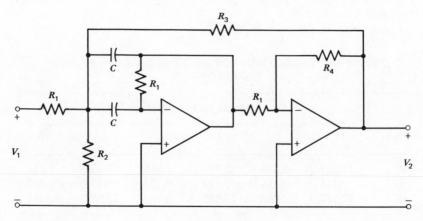

General circuit

Procedure

Given center frequency f_0 (hertz), gain G, and Q (or bandwidth $B = f_0/Q$), perform the following steps for a second-order filter, or for each identical stage of a higher-order cascaded filter.

1. Select a value of capacitance C and determine a K parameter from

$$K = \frac{100}{f_0 C'}$$

where C' is the value of C in microfarads. Alternatively, K may be found from Fig. 4-14a, b, or c.

2. Find the resistance values from the appropriate one of Tables 4-18 through 4.27. The resistances in the tables are given for $K = 1$ and hence their values must be multiplied by the K parameter of step 1 to yield the resistances of the circuit.

3. Select standard resistance values that are as close as possible to those indicated by the table and construct the filter, or its stages, in accordance with the general circuit.

Comments and Suggestions

The comments and suggestions for the VCVS bandpass filter given in Sec. 4.9 apply as follows:

(a) Paragraphs (b), (d), and (e) are directly applicable.

(b) Paragraphs (a) and (c) do not apply.

In addition, the following applies:

(c) The value of Q (and hence B) can be varied to some degree, without appreciably affecting f_0, by varying R_3 or R_4.

(d) For minimum dc offset, resistances equal to R_1 and $R_1 R_4/(R_1 + R_4)$ respectively, may be placed in the noninverting input leads of the op-amps.

A discussion of the PFB bandpass filter is given in Sec. 4.5.

Table 4-18. Second-Order Positive-Feedback Bandpass Filter Designs ($Q = 5$)

Gain	Circuit Element Values[a]					
	1	2	4	6	8	10
R_1	3.559	3.559	2.516	2.906	2.516	2.251
R_2	6.742	1.572	14.903	1.831	3.016	5.729
R_3	1.025	2.050	1.890	3.894	3.781	3.707
R_4	1.592	3.183	3.183	6.366	6.366	6.366

[a] Resistances in kilohms for a K parameter of 1.

Table 4-19. Second-Order Positive-Feedback Bandpass
Filter Designs ($Q = 10$)

	Circuit Element Values[a]					
Gain	1	2	4	6	8	10
R_1	5.033	5.033	3.559	2.906	3.559	3.183
R_2	1.369	0.794	1.767	4.310	1.184	1.516
R_3	0.945	1.890	1.792	1.751	3.584	3.537
R_4	1.592	3.183	3.183	3.183	6.366	6.366

[a] Resistances in kilohms for a K parameter of 1.

Table 4-20. Second-Order Positive-Feedback Bandpass
Filter Designs ($Q = 15$)

	Circuit Element Values[a]					
Gain	1	2	4	6	8	10
R_1	6.164	4.359	4.359	3.559	3.082	2.757
R_2	0.850	2.862	1.087	1.843	3.284	7.492
R_3	0.914	0.876	1.751	1.720	1.701	1.689
R_4	1.592	1.592	3.183	3.183	3.183	3.183

[a] Resistances in kilohms for a K parameter of 1.

Table 4-21. Second-Order Positive-Feedback Bandpass
Filter Designs ($Q = 20$)

	Circuit Element Values[a]					
Gain	1	2	4	6	8	10
R_1	7.118	5.033	5.033	4.109	3.559	3.183
R_2	0.644	1.585	0.827	1.264	1.884	2.894
R_3	0.896	0.864	1.728	1.701	1.686	1.675
R_4	1.592	1.592	3.183	3.183	3.183	3.183

[a] Resistances in kilohms for a K parameter of 1.

Table 4-22.   Second-Order Positive-Feedback Bandpass
Filter Designs ($Q = 25$)

Gain	Circuit Element Values[a]					
	1	2	4	6	8	10
R_1	7.958	5.627	5.627	4.594	3.979	3.559
R_2	0.531	1.142	0.685	0.996	1.384	1.909
R_3	0.884	0.856	1.713	1.689	1.675	1.666
R_4	1.592	1.592	3.183	3.183	3.183	3.183

[a] Resistances in kilohms for a K parameter of 1.

Table 4-23.   Second-Order Positive-Feedback Bandpass
Filter Designs ($Q = 30$)

Gain	Circuit Element Values[a]					
	1	2	4	6	8	10
R_1	7.797	5.513	6.164	5.033	4.359	3.898
R_2	0.663	1.941	0.594	0.838	1.122	1.471
R_3	0.693	0.676	1.701	1.680	1.668	1.659
R_4	1.273	1.273	3.183	3.183	3.183	3.183

[a] Resistances in kilohms for a K parameter of 1.

Table 4-24.   Second-Order Positive-Feedback Bandpass
Filter Designs ($Q = 35$)

Gain	Circuit Element Values[a]					
	1	2	4	6	8	10
R_1	7.688	5.436	6.658	5.436	4.708	4.211
R_2	0.869	5.911	0.530	0.733	0.957	1.219
R_3	0.570	0.558	1.693	1.673	1.662	1.654
R_4	1.061	1.061	3.183	3.183	3.183	3.183

[a] Resistances in kilohms for a K parameter of 1.

Table 4-25. Second-Order Positive-Feedback Bandpass
Filter Designs ($Q = 40$)

Gain	Circuit Element Values[a]					
	1	2	4	6	8	10
R_1	8.219	5.812	5.033	5.812	5.033	4.502
R_2	0.735	3.093	1.720	0.657	0.844	1.054
R_3	0.567	0.556	0.829	1.668	1.657	1.650
R_4	1.061	1.061	1.592	3.183	3.183	3.183

[a] Resistances in kilohms for a K parameter of 1.

Table 4-26. Second-Order Positive-Feedback Bandpass
Filter Designs ($Q = 45$)

Gain	Circuit Element Values[a]					
	1	2	4	6	8	10
R_1	8.071	7.549	5.338	4.359	5.338	4.775
R_2	1.013	0.603	1.408	3.664	0.760	0.936
R_3	0.482	0.840	0.827	0.821	1.653	1.646
R_4	0.909	1.592	1.592	1.592	3.183	3.183

[a] Resistances in kilohms for a K parameter of 1.

Table 4-27. Second-Order Positive-Feedback Bandpass
Filter Designs ($Q = 50$)

Gain	Circuit Element Values[a]					
	1	2	4	6	8	10
R_1	8.507	7.958	5.627	4.594	5.627	5.033
R_2	0.863	0.549	1.203	2.661	0.696	0.848
R_3	0.480	0.838	0.825	0.819	1.650	1.644
R_4	0.909	1.592	1.592	1.592	3.183	3.183

[a] Resistances in kilohms for a K parameter of 1.

4.12 Summary of Biquad Bandpass Filter Design Procedure ($Q \leq 100$)

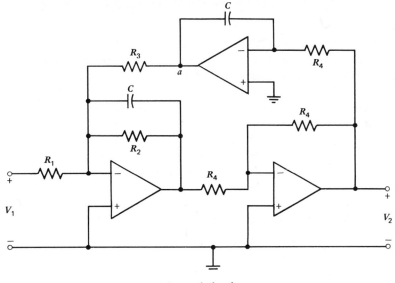

General circuit

Procedure

Given center frequency f_0 (hertz), gain G, order n, and filter type (Butter-worth or Chebyshev), perform the following steps for a second-order filter, or for each stage of a fourth-order cascaded filter.

1. Select a value of capacitance C and determine a K parameter from

$$K = \frac{100}{f_0 C'}$$

where C' is the value of C in microfarads. Alternately, K may be found from Fig. 4-14a, b, or c. For fourth-order designs, it is better to use the equation since greater accuracy is required.

2. Find the resistance values from the appropriate one of Tables 4-28 through 4-34. The resistances in the tables are given for $K = 1$ and hence their values must be multiplied by the K parameter of step 1 to yield the resistances of the circuit. The number G in the tables is the stage gain, and for fourth order, the filter gain is the product of the stage gains. The stage gains, which are not necessarily equal, are chosen by the designer.

3. Select standard resistance values that are as close as possible to those indicated by the table and construct the filter, or its stages, in accordance with the general circuit.

Comments and Suggestions

The comments and suggestions for the VCVS bandpass filter given in Sec. 4.9 apply as follows:

(a) Paragraphs (a) and (c) do not apply.

(b) Paragraph (b) applies with the exception that the resistors should be of 2% tolerance and the capacitors of at most 5% tolerance, in the fourth-order case.

(c) Paragraph (d) applies except that the statement concerning open-loop gain is not necessary.

In addition, the following applies:

(d) The stage gain is R_2/R_1. If an inverting gain of the same magnitude is desired, the output can be taken at point a.

(e) The filter response is readily tuned by varying R_1, R_2, and R_3. Varying R_1 affects the gain, varying R_2 affects Q, and varying R_3 changes f_0.

A specific example was given in Sec. 4.6.

Table 4-28. Second-Order Bandpass Biquad Filter Designs

Circuit Element Values[a]	
R_1	$1.592Q/G$
R_2	$1.592Q$
R_3, R_4	1.592

[a] Resistances in kilohms for a K parameter of 1, G = gain, and Q = quality factor of the filter.

Table 4-29. Fourth-Order Bandpass Butterworth Cascaded Biquad Filter Designs

Circuit Element Values[a]

Q	2	4	6	8	10	20	30	40	50	Stage
R_1	3.183/G	6.366/G	9.549/G	12.732/G	15.915/G	31.831/G	47.746/G	63.662/G	79.577/G	1
R_2	3.826	8.272	12.753	17.244	21.739	44.234	66.351	89.243	111.750	
R_3, R_4	1.114	1.333	1.414	1.457	1.483	1.536	1.536	1.564	1.569	
R_1	3.183/G	6.366/G	9.549/G	12.732/G	15.915/G	31.831/G	47.746/G	63.662/G	79.577/G	2
R_2	5.468	9.876	14.350	18.839	23.333	45.826	68.739	90.835	113.341	
R_3, R_4	2.275	1.900	1.791	1.739	1.708	1.649	1.649	1.620	1.614	

[a] Resistances in kilohms for a K parameter of 1. G = stage gain at the center frequency of the filter. The overall gain is the product of the stage gains.

Table 4-30. Fourth-Order Bandpass Chebyshev Cascaded Biquad Filter Designs (0.1 dB)

	Circuit Element Values[a]									Stage
Q	2	4	6	8	10	20	30	40	50	
R_1	1.749/G	3.497/G	5.246/G	6.995/G	8.744/G	17.487/G	26.231/G	34.974/G	43.718/G	1
R_2	2.005	4.580	7.222	9.883	12.553	25.943	39.351	52.765	66.181	
R_3, R_4	0.787	1.125	1.264	1.339	1.386	1.485	1.520	1.538	1.548	
R_1	1.749/G	3.497/G	5.246/G	6.995/G	8.744/G	17.487/G	26.231/G	34.974/G	43.718/G	2
R_2	4.056	6.482	9.096	11.748	14.413	27.797	41.205	54.619	68.034	
R_3, R_4	3.219	2.252	2.005	1.892	1.827	1.705	1.667	1.647	1.636	

[a] Resistances in kilohms for a K-parameter of 1. G = stage gain at the center frequency of the filter. The overall gain is the product of the stage gains.

164

Table 4-31. Fourth-Order Bandpass Chebyshev Cascaded Biquad Filter Designs (0.5 dB)

				Circuit Element Values[a]						
Q	2	4	6	8	10	20	30	40	50	Stage
R_1	2.585/G	5.170/G	7.755/G	10.340/G	12.925/G	25.851/G	38.776/G	51.701/G	64.627/G	1
R_2	3.581	7.939	12.364	16.808	21.261	43.562	65.881	88.204	110.529	
R_3, R_4	0.961	1.238	1.346	1.404	1.439	1.514	1.539	1.552	1.560	
R_1	2.585/G	5.170/G	7.755/G	10.340/G	12.925/G	25.851/G	38.776/G	51.701/G	64.627/G	2
R_2	5.930	10.207	14.618	19.057	23.507	45.805	68.123	90.446	112.771	
R_3, R_4	2.635	2.046	1.882	1.804	1.760	1.673	1.646	1.632	1.624	

[a] Resistances in kilohms for a K parameter of 1. G = stage gain at the center frequency of the filter. The overall gain is the product of the stage gains.

Table 4-32. Fourth-Order Bandpass Chebyshev Cascaded Biquad Filter Designs (1 dB)

Q	2	4	6	8	10	20	30	40	50	Stage
				Circuit Element Values[a]						
R_1	3.032/G	6.063/G	9.095/G	12.126/G	15.158/G	30.315/G	45.473/G	60.630/G	75.788/G	1
R_2	4.753	10.436	16.192	21.970	27.756	56.725	85.712	114.705	143.699	
R_3, R_4	1.017	1.272	1.371	1.423	1.455	1.522	1.545	1.556	1.563	
R_1	3.032/G	6.063/G	9.095/G	12.126/G	15.158/G	30.315/G	45.473/G	60.630/G	75.788/G	2
R_2	7.438	13.054	18.798	24.571	30.355	59.321	88.308	117.300	146.295	
R_3, R_4	2.491	1.991	1.848	1.780	1.741	1.664	1.640	1.628	1.620	

[a] Resistances in kilohms for a K parameter of 1. G = stage gain at the center frequency of the filter. The overall gain is the product of the stage gains.

Table 4-33. Fourth-Order Bandpass Chebyshev Cascaded Biquad Filter Designs (2 dB)

Q	2	4	6	8	10	20	30	40	50	Stage
				Circuit Element Values[a]						
R_1	3.509/G	7.017/G	10.526/G	14.035/G	17.544/G	35.087/G	52.631/G	70.175/G	87.718/G	1
R_2	6.599	14.383	22.254	30.149	38.053	77.621	117.210	156.805	196.401	
R_3, R_4	1.061	1.299	1.390	1.438	1.467	1.528	1.549	1.560	1.566	
R_1	3.509/G	7.017/G	10.526/G	14.035/G	17.544/G	35.087/G	52.631/G	70.175/G	87.718/G	2
R_2	9.902	17.625	25.484	33.375	41.277	80.843	120.431	160.025	199.622	
R_3, R_4	2.388	1.950	1.823	1.762	1.726	1.658	1.635	1.624	1.618	

[a] Resistances in kilohms for a K parameter of 1. G = stage gain at the center frequency of the filter. The overall gain is the product of the stage gains.

167

Table 4-34. Fourth-Order Bandpass Chebyshev Cascaded Biquad Filter Designs (3 dB)

Circuit Element Values[a]

Q	2	4	6	8	10	20	30	40	50	Stage
R_1	3.783/G	7.566/G	11.349/G	15.132/G	18.916/G	37.831/G	56.747/G	75.662/G	94.578/G	1
R_2	8.286	18.001	27.816	37.659	47.513	96.835	146.181	195.532	244.886	
R_3, R_4	1.080	1.311	1.398	1.444	1.473	1.531	1.551	1.561	1.567	
R_1	3.783/G	7.566/G	11.349/G	15.132/G	18.916/G	37.831/G	56.747/G	75.662/G	94.578/G	2
R_2	12.207	21.858	31.662	41.500	51.352	100.672	150.017	199.369	248.724	
R_3, R_4	2.345	1.933	1.812	1.754	1.720	1.655	1.633	1.623	1.616	

[a] Resistances in kilohms for a K parameter of 1. G = stage gain at the center frequency of the filter. The overall gain is the product of the stage gains.

4.13   Summary of Fourth-Order Multiple-Feedback Bandpass Filter Design Procedure ($Q \leq 10$)

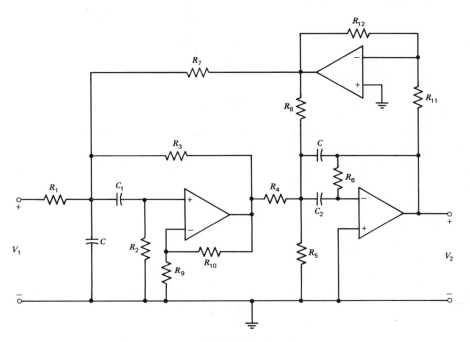

General circuit

Procedure

Given center frequency f_0 (hertz), gain G, and filter type (Butterworth or Chebyshev), perform the following steps.

1.   Select a value of capacitance C and determine a K parameter from

$$K = \frac{100}{f_0 C'}$$

where C' is the value of C in microfarads.

2.   Find the remaining element values from the appropriate one of Tables 4-35 through 4-58 as follows. The values of the capacitances other than C are determined directly from the tables using the chosen value of C. The resistances in the tables are given for $K = 1$ and hence their values must be multiplied by the K parameter of step 1 to yield the resistances of the circuit.

3.   Select standard resistance values that are as close as possible to those indicated by the table and construct the filter in accordance with the general

circuit. In case the remaining capacitances are multiples of C such as 0.47 and so forth, standard values result if C is chosen as a power of 10 (i.e., 0.1, 1, 10, etc.) μF.

Comments and Suggestions

The comments and suggestions for the VCVS bandpass filter given in Sec. 4.9 apply as follows:

(a) Paragraphs (a) and (c) apply except that R_4 and R_5 should be R_9 and R_{10}.

(b) Paragraph (b) applies except that the resistors should be of 2% tolerance and the capacitors of at most 5% tolerance.

(c) Paragraph (d) applies except for the statement concerning open-loop gain.

(d) Paragraph (e) does not apply.

A specific example was given in Sec. 4.7.

Table 4-35.   Fourth-Order Bandpass Butterworth
Multiple-Feedback Filter Designs ($Q = 1$)

Gain	Circuit Element Values[a]			
	1	2	6	10
R_1	4.522	4.862	5.160	3.113
R_2	9.115	11.021	12.802	18.977
R_3	3.404	4.168	5.155	4.048
R_4	1.347	0.680	0.268	0.236
R_5	0.269	0.223	0.526	0.392
R_6	2.142	2.040	1.769	1.709
R_7	9.773	10.211	5.481	8.563
R_8	40.135	29.378	3.283	6.313
R_9	15.603	17.865	18.416	28.974
R_{10}	21.922	28.771	42.000	54.999
R_{11}	10.000	10.000	10.000	10.000
R_{12}	70.271	55.942	10.902	22.502
C_1	C	C	C	C
C_2	C	C	C	C

[a] Resistances in kilohms for a K parameter of 1.

Table 4-36. Fourth-Order Bandpass Butterworth Multiple-Feedback Filter Designs ($Q = 2$)

Gain	Circuit Element Values[a]			
	1	2	6	10
R_1	12.285	6.215	3.200	2.419
R_2	8.021	9.756	11.835	13.898
R_3	2.749	2.505	2.910	2.958
R_4	1.743	1.720	1.383	1.215
R_5	0.263	0.259	0.246	0.237
R_6	2.267	2.265	2.259	2.278
R_7	6.944	6.851	7.119	6.760
R_8	15.040	14.998	11.280	9.179
R_9	15.222	18.548	19.134	21.209
R_{10}	16.954	20.581	31.022	40.321
R_{11}	10.000	10.000	10.000	10.000
R_{12}	76.349	76.125	57.064	45.562
C_1	C	C	C	C
C_2	C	C	C	C

[a] Resistances in kilohms for a K parameter of 1.

Table 4-37. Fourth-Order Bandpass Butterworth Multiple-Feedback Filter Designs ($Q = 3$)

Gain	Circuit Element Values[a]			
	1	2	6	10
R_1	25.898	13.648	6.059	3.656
R_2	8.310	8.680	8.885	11.625
R_3	2.421	2.459	3.111	2.659
R_4	1.701	1.669	1.591	1.544
R_5	0.260	0.259	0.253	0.238
R_6	2.272	2.272	2.300	2.335
R_7	6.591	6.748	6.694	6.523
R_8	12.867	12.565	9.683	10.123
R_9	17.224	17.376	14.665	19.505
R_{10}	16.057	17.346	22.540	28.775
R_{11}	10.000	10.000	10.000	10.000
R_{12}	81.087	79.145	59.914	61.174
C_1	C	C	C	C
C_2	C	C	C	C

[a] Resistances in kilohms for a K parameter of 1.

Table 4-38. Fourth-Order Bandpass Butterworth Multiple-Feedback Filter Designs ($Q = 4$)

Gain	Circuit Element Values[a]			
	1	2	6	10
R_1	28.801	25.064	8.148	5.970
R_2	7.773	7.502	9.806	9.805
R_3	1.910	2.658	2.253	2.685
R_4	2.499	1.707	1.648	1.566
R_5	0.340	0.269	0.262	0.262
R_6	2.233	2.283	2.271	2.210
R_7	6.403	6.803	6.825	6.734
R_8	12.354	10.725	11.495	9.458
R_9	17.790	14.254	19.729	17.311
R_{10}	13.803	15.839	19.495	22.612
R_{11}	10.000	10.000	10.000	10.000
R_{12}	88.402	73.947	79.663	67.763
C_1	C	C	C	C
C_2	C	C	C	C

[a] Resistances in kilohms for a K parameter of 1.

Table 4-39. Fourth-Order Bandpass Butterworth Multiple-Feedback Filter Designs ($Q = 5$)

Gain	Circuit Element Values[a]			
	1	2	6	10
R_1	61.539	31.580	12.534	9.025
R_2	7.779	7.903	8.784	8.558
R_3	2.196	2.268	2.354	2.792
R_4	1.907	1.924	1.684	1.609
R_5	0.285	0.272	0.266	0.266
R_6	2.272	2.335	2.272	2.224
R_7	6.904	6.673	6.886	6.858
R_8	12.082	11.874	10.959	9.251
R_9	16.895	16.506	17.571	15.178
R_{10}	14.416	15.165	17.564	19.620
R_{11}	10.000	10.000	10.000	10.000
R_{12}	88.456	84.109	80.046	69.293
C_1	C	C	C	C
C_2	C	C	C	C

[a] Resistances in kilohms for a K parameter of 1.

Table 4-40.   Fourth-Order Bandpass Butterworth
Multiple-Feedback Filter Designs ($Q = 10$)

Gain	Circuit Element Values[a]		
	2	6	10
R_1	584.854	199.060	78.575
R_2	17.249	17.396	15.143
R_3	3.530	3.417	2.711
R_4	0.449	0.436	0.633
R_5	0.375	0.432	0.670
R_6	2.265	2.231	2.204
R_7	17.470	16.373	16.345
R_8	10.394	8.835	7.744
R_9	33.352	33.894	30.837
R_{10}	35.725	35.737	29.753
R_{11}	10.000	10.000	10.000
R_{12}	61.708	53.624	47.641
C_1	0.3C	0.3C	0.3C
C_2	C	C	C

[a] Resistances in kilohms for a K parameter
of 1.

Table 4-41.   Fourth-Order Bandpass Chebyshev
Multiple-Feedback Filter Designs
(0.1 dB, $Q = 1$)

Gain	Circuit Element Values[a]			
	1	2	6	10
R_1	2.758	2.109	1.952	1.872
R_2	4.852	6.751	10.539	13.735
R_3	2.611	2.400	2.599	2.717
R_4	0.633	0.425	0.164	0.107
R_5	0.225	0.211	0.198	0.189
R_6	2.288	2.268	2.239	2.240
R_7	17.513	16.561	18.216	18.732
R_8	7.060	7.124	7.381	7.428
R_9	8.629	11.780	17.504	22.181
R_{10}	11.083	15.812	16.490	36.073
R_{11}	10.000	10.000	10.000	10.000
R_{12}	40.933	39.725	37.478	35.811
C_1	3C	3C	3C	3C
C_2	C	C	C	C

[a] Resistances in kilohms for a K parameter of 1.

Table 4-42.   Fourth-Order Bandpass Chebyshev
Multiple-Feedback Filter Designs
(0.1 dB, $Q = 2$)

Gain	Circuit Element Values[a]			
	1	2	6	10
R_1	15.784	7.352	3.810	3.242
R_2	2.943	3.310	4.877	6.391
R_3	2.342	2.239	2.387	2.463
R_4	0.405	0.433	0.295	0.212
R_5	0.367	0.373	0.332	0.294
R_6	2.268	2.267	2.235	2.229
R_7	15.831	15.738	15.759	15.896
R_8	5.931	5.906	5.727	5.764
R_9	5.648	6.371	8.917	11.528
R_{10}	6.145	6.891	10.765	14.343
R_{11}	10.000	10.000	10.000	10.000
R_{12}	44.384	44.493	42.383	41.701
C_1	3C	3C	3C	3C
C_2	C	C	C	C

[a] Resistances in kilohms for a K parameter of 1.

Table 4-43. Fourth-Order Bandpass Chebyshev
Multiple-Feedback Filter Designs
(0.1 dB, $Q = 3$)

Gain	Circuit Element Values[a]			
	1	2	6	10
R_1	75.960	35.919	7.233	7.126
R_2	4.423	4.477	4.772	6.032
R_3	3.438	3.362	2.664	3.120
R_4	0.204	0.216	0.346	0.217
R_5	0.335	0.346	0.423	0.346
R_6	2.255	2.213	2.214	2.214
R_7	16.034	16.073	15.824	15.985
R_8	5.933	5.922	5.924	5.919
R_9	7.949	8.048	8.807	10.874
R_{10}	9.969	10.092	10.414	13.550
R_{11}	10.000	10.000	10.000	10.000
R_{12}	41.810	41.890	43.067	41.852
C_1	2C	2C	2C	2C
C_2	C	C	C	C

[a] Resistances in kilohms for a K parameter of 1.

Table 4-44.   Fourth-Order Bandpass Chebyshev
Multiple-Feedback Filter Designs
(0.1 dB, $Q = 4$)

Gain	Circuit Element Values[a]			
	1	2	6	10
R_1	188.432	66.373	23.131	16.567
R_2	5.064	4.480	4.939	5.627
R_3	3.828	3.309	3.335	3.547
R_4	0.148	0.206	0.198	0.167
R_5	0.293	0.353	0.347	0.316
R_6	2.212	2.212	2.212	2.213
R_7	16.060	16.019	16.043	16.070
R_8	5.533	5.572	5.567	5.546
R_9	9.023	8.101	8.920	10.076
R_{10}	11.543	10.025	11.069	12.743
R_{11}	10.000	10.000	10.000	10.000
R_{12}	41.355	42.134	42.031	41.601
C_1	2C	2C	2C	2C
C_2	C	C	C	C

[a] Resistances in kilohms for a K parameter of 1.

Table 4-45. Fourth-Order Bandpass Chebyshev
Multiple-Feedback Filter Designs
(0.1 dB, $Q = 5$)

Gain	\multicolumn{4}{c}{Circuit Element Values[a]}			
	1	2	6	10
R_1	162.467	125.704	45.556	27.218
R_2	3.986	4.806	5.175	5.421
R_3	2.949	3.510	3.632	3.599
R_4	0.259	0.171	0.158	0.159
R_5	0.411	0.330	0.316	0.315
R_6	2.212	2.212	2.211	2.212
R_7	16.052	16.114	16.141	16.120
R_8	5.392	5.357	5.330	5.342
R_9	7.299	8.658	9.263	9.722
R_{10}	8.782	10.800	11.725	12.254
R_{11}	10.000	10.000	10.000	10.000
R_{12}	42.704	41.892	41.603	41.685
C_1	2C	2C	2C	2C
C_2	C	C	C	C

[a] Resistances in kilohms for a K parameter of 1.

Table 4-46.   Fourth-Order Bandpass Chebyshev
Multiple-Feedback Filter Designs
(0.1 dB, $Q = 10$)

Gain	Circuit Element Values[a]		
	2	6	10
R_1	415.648	20.759	76.788
R_2	4.494	2.617	4.521
R_3	3.141	1.656	3.012
R_4	0.202	1.258	0.216
R_5	0.362	0.760	0.372
R_6	2.231	2.234	2.230
R_7	16.681	16.461	16.702
R_8	5.308	5.575	5.403
R_9	8.251	5.109	8.383
R_{10}	9.868	5.363	9.812
R_{11}	10.000	10.000	10.000
R_{12}	44.450	48.442	45.307
C_1	2C	2C	2C
C_2	C	C	C

[a] Resistances in kilohms for a K parameter
of 1.

Table 4-47. Fourth-Order Bandpass Chebyshev
Multiple-Feedback Filter Designs
$(0.5 \text{ dB}, Q = 1)$

Gain	Circuit Element Values[a]			
	1	2	6	10
R_1	1.049	1.231	7.484	3.898
R_2	13.935	12.851	7.285	15.171
R_3	1.581	1.546	7.867	5.913
R_4	3.965	1.491	0.161	0.140
R_5	0.107	0.108	3.792	1.722
R_6	3.829	3.857	1.783	1.678
R_7	9.805	17.427	36.967	19.948
R_8	5.334	6.191	11.229	8.882
R_9	23.292	23.580	9.474	21.893
R_{10}	34.686	28.243	31.532	49.405
R_{11}	10.000	10.000	10.000	10.000
R_{12}	10.710	11.830	10.658	10.723
C_1	C	C	0.5C	0.5C
C_2	2C	2C	2C	2C

[a] Resistances in kilohms for a K parameter of 1.

Table 4-48. Fourth-Order Bandpass Chebyshev
Multiple-Feedback Filter Designs
(0.5 dB, $Q = 2$)

Gain	Circuit Element Values[a]			
	1	2	6	10
R_1	2.440	3.028	3.319	3.492
R_2	7.293	6.655	6.921	9.338
R_3	1.344	1.432	1.488	1.651
R_4	5.149	1.959	0.566	0.330
R_5	0.190	0.173	0.184	0.113
R_6	3.364	3.567	3.470	3.903
R_7	5.787	10.547	37.653	22.441
R_8	3.886	4.896	10.254	11.716
R_9	15.576	15.240	16.988	22.228
R_{10}	13.713	11.814	11.679	16.103
R_{11}	10.000	10.000	10.000	10.000
R_{12}	16.464	18.642	37.220	31.231
C_1	C	C	C	C
C_2	2C	2C	2C	2C

[a] Resistances in kilohms for a K parameter of 1.

Table 4-49. Fourth-Order Bandpass Chebyshev
Multiple-Feedback Filter Designs
$(0.5 \text{ dB}, \ Q = 3)$

Gain	Circuit Element Values[a]			
	1	2	6	10
R_1	2.741	2.411	2.267	2.024
R_2	6.422	7.441	8.360	10.592
R_3	1.427	1.357	1.469	1.482
R_4	11.936	6.658	2.366	1.646
R_5	0.277	0.255	0.215	0.180
R_6	2.799	2.771	2.778	2.761
R_7	3.168	3.333	4.764	4.314
R_8	2.517	2.666	2.976	3.024
R_9	11.885	13.995	15.696	19.300
R_{10}	13.971	15.891	17.886	23.476
R_{11}	10.000	10.000	10.000	10.000
R_{12}	13.970	14.365	15.045	14.748
C_1	C	C	C	C
C_2	2C	2C	2C	2C

[a] Resistances in kilohms for a K parameter of 1.

Table 4-50. Fourth-Order Bandpass Chebyshev
Multiple-Feedback Filter Designs
$(0.5 \text{ dB}, Q = 4)$

Gain	Circuit Element Values[a]			
	1	2	6	10
R_1	5.908	4.336	2.975	2.238
R_2	6.370	5.637	6.205	6.749
R_3	1.554	1.639	1.899	2.075
R_4	10.301	6.891	3.735	3.329
R_5	0.278	0.289	0.256	0.247
R_6	2.597	2.556	2.565	2.562
R_7	2.214	3.546	4.458	4.755
R_8	2.704	2.870	2.809	2.681
R_9	11.358	10.198	10.360	10.526
R_{10}	14.505	12.603	15.475	18.809
R_{11}	10.000	10.000	10.000	10.000
R_{12}	15.647	16.885	15.987	15.165
C_1	C	C	C	C
C_2	2C	2C	2C	2C

[a] Resistances in kilohms for a K parameter of 1.

Table 4-51. Fourth-Order Bandpass Chebyshev
Multiple-Feedback Filter Designs
(0.5 dB, $Q = 5$)

Gain	Circuit Element Values[a]			
	1	2	6	10
R_1	14.871	5.865	3.190	2.169
R_2	5.298	5.383	5.927	6.733
R_3	1.727	1.805	1.986	2.136
R_4	6.064	7.535	5.727	5.717
R_5	0.271	0.262	0.270	0.261
R_6	2.456	2.277	2.464	2.461
R_7	3.269	6.817	4.063	4.166
R_8	3.578	11.867	3.492	3.481
R_9	9.867	10.206	9.577	10.153
R_{10}	11.440	11.390	15.555	19.990
R_{11}	10.000	10.000	10.000	10.000
R_{12}	21.494	75.870	20.898	20.752
C_1	C	C	C	C
C_2	2C	2C	2C	2C

[a] Resistances in kilohms for a K parameter of 1.

Table 4-52. Fourth-Order Bandpass Chebyshev Multiple-Feedback Filter Designs (0.5 dB, $Q = 10$)

Gain	Circuit Element Values[a]		
	2	6	10
R_1	406.471	132.119	66.445
R_2	17.542	17.627	15.790
R_3	3.577	3.409	3.005
R_4	0.426	0.432	0.512
R_5	0.371	0.431	0.631
R_6	2.266	2.232	2.209
R_7	17.587	16.383	16.409
R_8	10.345	8.867	7.402
R_9	33.917	34.429	31.027
R_{10}	36.336	36.118	32.153
R_{11}	10.000	10.000	10.000
R_{12}	61.134	53.596	45.324
C_1	0.3C	0.3C	0.3C
C_2	C	C	C

[a] Resistances in kilohms for a K parameter of 1.

Table 4-53. Fourth-Order Bandpass Chebyshev Multiple-Feedback Filter Designs
(1 dB, $Q = 1$)

Gain	Circuit Element Values[a]			
	1	2	6	10
R_1	7.534	5.221	4.713	4.784
R_2	6.095	7.697	11.119	13.699
R_3	3.212	3.306	4.147	4.799
R_4	0.699	0.514	0.207	0.128
R_5	0.444	0.435	0.464	0.622
R_6	2.199	2.170	2.171	2.152
R_7	16.484	16.795	19.382	20.470
R_8	11.968	11.760	11.957	11.438
R_9	10.816	13.451	18.317	21.876
R_{10}	13.963	17.995	28.294	36.655
R_{11}	10.000	10.000	10.000	10.000
R_{12}	40.625	39.775	36.167	33.652
C_1	C	C	C	C
C_2	C	C	C	C

[a] Resistances in kilohms for a K parameter of 1.

Table 4-54.   Fourth-Order Bandpass Chebyshev
Multiple-Feedback Filter Designs
$(1 \text{ dB}, Q = 2)$

Gain	Circuit Element Values[a]			
	1	2	6	10
R_1	78.639	40.658	12.875	8.019
R_2	6.431	6.854	7.866	9.211
R_3	3.803	3.827	3.655	3.539
R_4	0.270	0.259	0.273	0.262
R_5	0.413	0.401	0.409	0.400
R_6	2.212	2.212	2.212	2.211
R_7	16.328	16.251	16.185	16.117
R_8	6.811	6.866	6.851	6.856
R_9	11.339	12.142	13.958	16.354
R_{10}	14.857	15.736	18.023	21.091
R_{11}	10.000	10.000	10.000	10.000
R_{12}	40.856	41.100	41.105	41.069
C_1	C	C	C	C
C_2	C	C	C	C

[a] Resistances in kilohms for a K parameter of 1.

Table 4-55. Fourth-Order Bandpass Chebyshev
Multiple-Feedback Filter Designs
(1 dB, $Q = 3$)

Gain	Circuit Element Values[a]			
	1	2	6	10
R_1	270.356	124.860	43.315	22.242
R_2	8.411	7.751	8.370	8.298
R_3	4.166	4.167	4.194	3.827
R_4	0.167	0.189	0.181	0.208
R_5	0.329	0.350	0.342	0.399
R_6	2.213	2.211	2.211	2.202
R_7	16.064	16.075	16.056	16.524
R_8	6.179	5.930	5.941	5.998
R_9	15.512	13.808	14.938	15.003
R_{10}	18.375	17.668	19.035	18.567
R_{11}	10.000	10.000	10.000	10.000
R_{12}	42.840	41.272	41.311	42.072
C_1	C	C	C	C
C_2	C	C	C	C

[a] Resistances in kilohms for a K parameter of 1.

Table 4-56. Fourth-Order Bandpass Chebyshev
Multiple-Feedback Filter Designs
(1 dB, $Q = 4$)

Gain	Circuit Element Values[a]			
	1	2	6	10
R_1	569.326	284.742	77.921	57.394
R_2	8.660	8.741	8.379	9.226
R_3	4.544	4.540	4.079	4.529
R_4	0.147	0.147	0.174	0.146
R_5	0.305	0.305	0.354	0.304
R_6	2.211	2.211	2.206	2.211
R_7	16.086	16.075	16.282	16.076
R_8	5.540	5.544	5.662	5.529
R_9	15.425	15.585	15.276	16.417
R_{10}	19.744	19.905	18.557	21.062
R_{11}	10.000	10.000	10.000	10.000
R_{12}	41.290	41.319	42.433	41.208
C_1	C	C	C	C
C_2	C	C	C	C

[a] Resistances in kilohms for a K parameter of 1.

Table 4-57. Fourth-Order Bandpass Chebyshev
Multiple-Feedback Filter Designs
(1 dB, $Q = 5$)

Gain	Circuit Element Values[a]			
	1	2	6	10
R_1	864.692	419.459	138.962	86.747
R_2	8.591	8.522	8.666	8.951
R_3	4.449	4.388	4.367	4.436
R_4	0.151	0.155	0.156	0.150
R_5	0.312	0.318	0.318	0.311
R_6	2.212	2.212	2.212	2.211
R_7	16.155	16.155	16.118	16.113
R_8	5.360	5.369	5.366	5.344
R_9	15.366	15.267	15.535	15.999
R_{10}	19.487	19.289	19.601	20.319
R_{11}	10.000	10.000	10.000	10.000
R_{12}	41.655	41.731	41.713	41.538
C_1	C	C	C	C
C_2	C	C	C	C

[a] Resistances in kilohms for a K parameter of 1.

Table 4-58. Fourth-Order Bandpass Chebyshev
Multiple-Feedback Filter Designs
(1 dB, $Q = 10$)

Gain	Circuit Element Values[a]		
	6	10	20
R_1	295.988	232.683	143.686
R_2	7.041	7.711	8.248
R_3	3.260	3.612	4.019
R_4	0.272	0.212	0.178
R_5	0.392	0.368	0.345
R_6	2.251	2.237	2.214
R_7	17.104	16.781	16.292
R_8	5.948	5.530	5.110
R_9	13.437	14.459	14.984
R_{10}	14.794	16.521	18.348
R_{11}	10.000	10.000	10.000
R_{12}	49.229	46.011	42.953
C_1	C	C	C
C_2	C	C	C

[a] Resistances in kilohms for a K parameter
of 1.

4.14   Summary of Multiple-Resonator Bandpass Filter Design Procedure
(Order = 4, 6, 8)

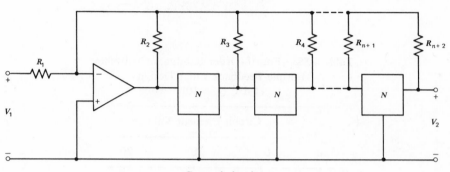

General circuit

Procedure

Given center frequency f_0 (hertz), gain G, order $2n$, and filter type (Butterworth or Chebyshev), perform the following steps.

1.   From the appropriate one of Tables 4-59 through 4-61, obtain the values of the resistances $R_1, R_2, \ldots, R_{n+2}$, of the feedback circuit, and the gain K_1 and quality factor Q_1 required of each of the n identical resonators N. (Recall that the order of the filter is $2n$.)

2.   Design the resonators N with noninverting gain K_1, quality factor Q_1, and center frequency f_0. This may be done readily using the biquad bandpass filter design of Sec. 4.12. If $Q \le 10$, in many cases, the PFB bandpass filter design of Sec. 4.11 is a suitable resonator. In this latter case, Tables 4-62 through 4-79 yield the resonator design, rather than Tables 4-18 through 4-27.

Comments and Suggestions

(a) The resistances $R_1, R_2, \ldots, R_{n+2}$ may all be multiplied by any convenient scale factor. There is no K parameter involved in the calculation of these resistances.

(b) Each resonator may be tuned separately to the desired f_0 of the filter, and the required K_1 and Q_1.

(c) The filter gain may be adjusted by varying R_1.

(d) The passband response is tuned by varying the feedback resistances $R_2, R_3, \ldots, R_{n+2}$.

(e) For fourth-, sixth-, and eighth-order filters, the resistances should be of 5%, 2%, and 1% tolerances, respectively. In the case of capacitors, percentage tolerances should parallel those of the resistors.

A specific example of a multiple-resonator bandpass filter was given in Sec. 4.8.

Table 4-59. Fourth-Order Bandpass Multiple-Resonator Filter Designs

	Circuit Element Values[a]					
		Chebyshev				
	Butterworth	0.1 dB	0.5 dB	1 dB	2 dB	3 dB
R_1	10.000/G	3.018/G	6.595/G	9.070/G	12.151/G	14.125/G
R_2	10.000	10.000	10.000	10.000	10.000	10.000
R_3	14.142	8.431	14.029	18.219	24.881	31.013
R_4	16.000	4.428	8.810	11.408	14.248	15.874
K_1	2.828	1.686	2.806	3.644	4.976	6.203
Q_1	2.828Q	1.686Q	2.806Q	3.644Q	4.976Q	6.203Q

[a] Resistances in kilohms and may all be multiplied by a convenient scale factor. G and Q are the gain and quality factor of the filter, and K_1 and Q_1 are the gain and quality factor required for each of the two resonators.

Table 4-60. Sixth-Order Bandpass Multiple-Resonator Filter Designs

	Circuit Element Values[a]					
		Chebyshev				
	Butterworth	0.1 dB	0.5 dB	1 dB	2 dB	3 dB
R_1	10.000/G	6.105/G	13.972/G	20.354/G	30.591/G	39.905/G
R_2	10.000	10.000	10.000	10.000	10.000	10.000
R_3	10.000	10.316	15.963	20.236	27.107	33.487
R_4	10.000	5.918	8.753	10.058	11.285	11.916
R_5	19.286	10.448	22.691	32.294	47.508	61.305
K_1	3.000	3.095	4.789	6.071	8.132	10.046
Q_1	3.000Q	3.095Q	4.789Q	6.071Q	8.132Q	10.046Q

[a] Resistances in kilohms and may all be multiplied by a convenient scale factor. G and Q are the gain and quality factor of the filter, and K_1 and Q_1 are the gain and quality factor required for each of the three resonators.

Table 4-61. Eighth-Order Bandpass Multiple-Resonator Filter Designs

Circuit Element Values[a]

	Butterworth	Chebyshev				
		0.1 dB	0.5 dB	1 dB	2 dB	3 dB
R_1	$10.000/G$	$12.070/G$	$26.382/G$	$36.281/G$	$48.599/G$	$56.501/G$
R_2	10.000	10.000	10.000	10.000	10.000	10.000
R_3	7.654	11.088	16.703	20.991	27.925	34.389
R_4	6.695	5.842	7.612	8.343	8.991	9.311
R_5	9.262	9.344	17.284	23.251	32.661	41.243
R_6	23.200	20.523	38.385	48.451	59.130	65.128
K_1	3.061	4.435	6.681	8.396	11.170	13.756
Q_1	$3.061Q$	$4.435Q$	$6.681Q$	$8.396Q$	$11.170Q$	$13.756Q$

[a] Resistances in kilohms and may all be multiplied by a convenient scale factor. G and Q are the gain and quality factor of the filter, and K_1 and Q_1 are the gain and quality factor required for each of the four resonators.

Table 4-62. PFB Resonator Designs for Fourth-Order
Multiple-Resonator Butterworth Bandpass Filters

Circuit Element Values[a]

Q	1	2	4	6	8	10
K_1	2.828	2.828	2.828	2.828	2.828	2.828
Q_1	2.828	5.656	11.312	16.968	22.624	28.280
R_1	3.183	4.502	7.118	8.717	4.502	5.033
R_2	1.352	0.758	0.405	0.321	2.653	1.662
R_3	4.924	4.244	4.959	4.745	0.849	0.843
R_4	6.366	6.366	7.958	7.958	1.592	1.592

[a] Resistances in kilohms for a K parameter of 1.

Table 4-63. PFB Resonator Designs for Fourth-Order Multiple-Resonator
Chebyshev Bandpass Filters (0.1 dB)

Q	1	2	4	6	8	10
			Circuit Element Values[a]			
K_1	1.686	1.686	1.686	1.686	1.686	1.686
Q_1	1.686	3.372	6.744	10.116	13.488	16.860
R_1	3.183	4.502	6.366	7.797	9.003	10.066
R_2	1.228	0.729	0.468	0.369	0.313	0.276
R_3	7.823	5.482	4.525	4.200	4.028	3.918
R_4	6.366	6.366	6.366	6.366	6.366	6.366

[a] Resistances in kilohms for a K parameter of 1.

Table 4-64. PFB Resonator Designs for Fourth-Order
Multiple-Resonator Chebyshev Filters (0.5 dB)

Q	1	2	4	6	8	10
			Circuit Element Values[a]			
K_1	2.806	2.806	2.806	2.806	2.806	2.806
Q_1	2.806	5.612	11.224	16.836	22.448	28.060
R_1	3.183	4.502	6.366	7.797	9.003	10.066
R_2	1.351	0.758	0.477	0.373	0.316	0.278
R_3	4.946	4.255	3.873	3.725	3.642	3.587
R_4	6.366	6.366	6.366	6.366	6.366	6.366

[a] Resistances in kilohms for a K parameter of 1.

Table 4-65. PFB Resonator Designs for Fourth-Order
Multiple-Resonator Chebyshev Bandpass Filters (1 dB)

	Circuit Element Values[a]					
Q	1	2	4	6	8	10
K_1	3.644	3.644	3.644	3.644	3.644	3.644
Q_1	3.644	7.288	14.576	21.864	29.152	36.440
R_1	3.183	4.502	6.366	7.797	9.003	10.066
R_2	1.400	0.768	0.480	0.374	0.317	0.279
R_3	4.387	3.949	3.689	3.585	3.525	3.486
R_4	6.366	6.366	6.366	6.366	6.366	6.366

[a] Resistances in kilohms for a K parameter of 1.

Table 4-66. PFB Resonator Designs for Fourth-Order
Multiple Resonator Chebyshev Bandpass Filters (2 dB)

	Circuit Element Values[a]					
Q	1	2	4	6	8	10
K_1	4.976	4.976	4.976	4.976	4.976	4.976
Q_1	4.976	9.952	19.904	29.856	39.808	49.760
R_1	3.183	4.502	4.502	3.898	4.502	5.033
R_2	1.446	0.778	1.030	12.910	2.915	1.750
R_3	3.984	3.710	1.713	0.830	0.825	0.822
R_4	6.366	6.366	3.183	1.592	1.592	1.592

[a] Resistances in kilohms for a K parameter of 1.

Table 4-67. PFB Resonator Designs for Fourth-Order Multiple-Resonator Chebyshev Bandpass Filters (3 dB)

Q	Circuit Element Values[a]					
	1	2	4	6	8	10
K_1	6.203	6.203	6.203	6.203	6.203	6.203
Q_1	6.203	12.406	24.812	37.218	49.624	62.030
R_1	3.183	3.183	4.502	5.513	4.502	7.118
R_2	1.473	2.741	1.039	0.716	2.992	0.485
R_3	3.795	1.731	1.688	1.669	0.819	1.651
R_4	6.366	3.183	3.183	3.183	1.592	3.183

[a] Resistances in kilohms for a K parameter of 1.

Table 4-68. PFB Resonator Designs for Sixth-Order Multiple-Resonator Butterworth Bandpass Filters

Q	Circuit Element Values[a]					
	1	2	4	6	8	10
K_1	3.000	3.000	3.000	3.000	3.000	3.000
Q_1	3.000	6.000	12.000	18.000	24.000	30.000
R_1	3.183	3.183	4.502	5.513	5.198	5.812
R_2	1.364	2.387	0.999	0.701	1.019	0.808
R_3	4.775	1.910	1.804	1.761	1.138	1.130
R_4	6.366	3.183	3.183	3.183	2.122	2.122

[a] Resistances in kilohms for a K parameter of 1.

Table 4-69. PFB Resonator Designs for Sixth-Order
Multiple-Resonator Chebyshev Bandpass Filters (0.1 dB)

	Circuit Element Values[a]					
Q	1	2	4	6	8	10
K_1	3.095	3.095	3.095	3.095	3.095	3.095
Q_1	3.095	6.190	12.380	18.570	24.760	30.950
R_1	3.183	4.502	4.502	5.513	6.366	7.118
R_2	1.370	0.762	1.002	0.702	0.562	0.479
R_3	4.702	4.126	1.797	1.755	1.731	1.715
R_4	6.366	6.366	3.183	3.183	3.183	3.183

[a] Resistances in kilohms for a K parameter of 1.

Table 4-70. PFB Resonator Designs for Sixth-Order
Multiple-Resonator Chebyshev Bandpass Filters (0.5 dB)

	Circuit Element Values[a]					
Q	1	2	4	6	8	10
K_1	4.789	4.789	4.789	4.789	4.789	4.789
Q_1	4.789	9.578	19.156	28.734	38.312	47.890
R_1	3.183	4.502	4.502	5.513	6.366	7.118
R_2	1.441	0.777	1.028	0.712	0.568	0.483
R_3	4.023	3.735	1.718	1.694	1.679	1.670
R_4	6.366	6.366	3.183	3.183	3.183	3.183

[a] Resistances in kilohms for a K parameter of 1.

Table 4-71.   PFB Resonator Designs for Sixth-Order
Multiple-Resonator Chebyshev Bandpass Filters (1 dB)

Q	Circuit Element Values[a]					
	1	2	4	6	8	10
K_1	6.071	6.071	6.071	6.071	6.071	6.071
Q_1	6.071	12.142	24.284	36.426	48.568	60.710
R_1	3.183	4.502	4.502	5.513	6.366	5.033
R_2	1.470	0.783	1.038	0.716	0.570	1.772
R_3	3.811	3.603	1.690	1.671	1.660	0.817
R_4	6.366	6.366	3.183	3.183	3.183	1.592

[a] Resistances in kilohms for a K parameter of 1.

Table 4-72.   PFB Resonator Designs for Sixth-Order
Multiple-Resonator Chebyshev Bandpass Filters (2 dB)

Q	Circuit Element Values[a]				
	1	2	4	6	8
K_1	8.132	8.132	8.132	8.132	8.132
Q_1	8.132	16.264	32.528	48.792	65.056
R_1	3.183	4.502	4.502	5.513	6.366
R_2	1.499	0.789	1.048	0.720	0.572
R_3	3.629	3.486	1.664	1.650	1.642
R_4	6.366	6.366	3.183	3.183	3.183

[a] Resistances in kilohms for a K parameter of 1.

Table 4-73. PFB Resonator Designs for Sixth-Order
Multiple-Resonator Chebyshev Bandpass Filters
(3 dB)

	Circuit Element Values[a]			
Q	1	2	4	6
K_1	10.046	10.046	10.046	10.046
Q_1	10.046	20.092	40.184	60.276
R_1	3.183	4.502	4.502	5.513
R_2	1.516	0.792	1.054	0.722
R_3	3.535	3.424	1.650	1.639
R_4	6.366	6.366	3.183	3.183

[a] Resistances in kilohms for a K parameter of 1.

Table 4-74. PFB Resonator Designs for Eighth-Order
Multiple-Resonator Butterworth Bandpass Filters

	Circuit Element Values[a]					
Q	1	2	4	6	8	10
K_1	3.061	3.061	3.061	3.061	3.061	3.061
Q_1	3.061	6.122	12.244	18.366	24.488	30.610
R_1	3.183	4.502	4.502	5.513	4.502	5.033
R_2	1.368	0.761	1.001	0.701	2.696	1.676
R_3	4.728	4.139	1.799	1.757	0.845	0.839
R_4	6.366	6.366	3.183	3.183	1.592	1.592

[a] Resistances in kilohms for a K parameter of 1.

Table 4-75. PFB Resonator Designs for Eighth-Order
Multiple-Resonator Chebyshev Bandpass Fiiters (0.1 dB)

	Circuit Element Values[a]					
Q	1	2	4	6	8	10
K_1	4.435	4.435	4.435	4.435	4.435	4.435
Q_1	4.435	8.870	17.740	26.610	35.480	44.350
R_1	3.183	4.502	4.502	5.513	4.502	5.033
R_2	1.430	0.775	1.024	0.710	2.870	1.735
R_3	4.110	3.787	1.729	1.702	0.829	0.825
R_4	6.366	6.366	3.183	3.183	1.592	1.592

[a] Resistances in kilohms for a K parameter of 1.

Table 4-76. PFB Resonator Designs for Eighth-Order
Multiple-Resonator Chebyshev Bandpass Filters (0.5 dB)

	Circuit Element Values[a]					
Q	1	2	4	6	8	10
K_1	6.681	6.681	6.681	6.681	6.681	6.681
Q_1	6.681	13.362	26.724	40.086	53.448	66.810
R_1	3.183	4.502	4.502	5.513	4.502	5.033
R_2	1.481	0.785	1.042	0.717	3.015	1.781
R_3	3.743	3.560	1.680	1.663	0.817	0.815
R_4	6.366	6.366	3.183	3.183	1.592	1.592

[a] Resistances in kilohms for a K parameter of 1.

Table 4-77. PFB Resonator Designs for Eighth-Order
Multiple-Resonator Chebyshev Bandpass Filters (1 dB)

	Circuit Element Values[a]					
Q	1	2	3	4	5	6
K_1	8.396	8.396	8.396	8.396	8.396	8.396
Q_1	8.396	16.792	25.188	33.584	41.980	50.376
R_1	5.033	7.118	3.898	4.502	5.033	5.513
R_2	0.593	0.391	1.460	1.049	0.845	0.720
R_3	9.804	9.180	1.673	1.662	1.654	1.648
R_4	15.915	15.915	3.183	3.183	3.183	3.183

[a] Resistances in kilohms for a K parameter of 1.

Table 4-78. PFB Resonator Designs for Eighth-Order
Multiple-Resonator Chebyshev Bandpass Filters (2 dB)

	Circuit Element Values[a]				
Q	1	2	3	4	5
K_1	11.170	11.170	11.170	11.170	11.170
Q_1	11.170	22.340	33.510	44.680	55.850
R_1	5.033	7.118	3.898	4.502	5.033
R_2	0.595	0.391	1.477	1.056	0.849
R_3	9.270	8.843	1.652	1.644	1.638
R_4	15.915	15.915	3.183	3.183	3.183

[a] Resistances in kilohms for a K parameter of 1.

Table 4-79. PFB Resonator Designs for Eighth-Order
Multiple-Resonator Chebyshev Bandpass Filters (3 dB)

Q	Circuit Element Values[a]				
	1	2	3	4	5
K_1	13.756	13.756	13.756	13.756	13.756
Q_1	13.756	27.512	41.268	55.024	68.780
R_1	5.033	5.033	3.898	4.502	5.033
R_2	0.596	0.645	1.486	1.061	0.852
R_3	8.991	4.221	1.640	1.634	1.629
R_4	15.915	7.958	3.183	3.183	3.183

[a] Resistances in kilohms for a K parameter of 1.

5

Band-Reject Filters

5.1 General Theory

A *band-reject* filter (also called *band-elimination* or *notch*) is one that passes all frequencies except a single band. The amplitude response of a band-reject filter is shown in Fig. 5-1, where the broken line represents the ideal response and the solid line represents a realizable approximation to the ideal. The band of frequencies that is rejected is centered approximately at ω_0 and its width is B. We may measure B in radians/second or in hertz and consider the *center frequency* as ω_0 rad/sec or as $f_0 = \omega_0/2\pi$ Hz. As in the case of the bandpass filter, we define the quantity $Q = \omega_0/B$ (or f_0/B if B is in hertz). Thus, a large Q indicates a narrow band rejected and a small Q indicates a wide band. The stopband is the band rejected and the passband is the rest of the frequency spectrum. The *gain G* is the amplitude of the transfer function at zero or infinite frequency.

Also like the bandpass filter, the band-reject filter is characterized by two *cutoff points* ω_1 and ω_2, as shown in Fig. 5-1. The cutoff points are defined to occur at points where $|H(j\omega)|$ attains $1/\sqrt{2}$ times its maximum value. For the realizable approximation the bandwidth is $B = \omega_2 - \omega_1$.

As for the high-pass and bandpass filters, the band-reject filter transfer function may be obtained by transforming that of a low-pass prototype. The band-reject transfer function obtained in this way is given in the general case by [22]

$$\frac{V_2}{V_1} = \frac{Gb_0}{S^n + b_{n-1}S^{n-1} + \cdots + b_1 S + b_0}\bigg|_{S = Bs/(s^2 + \omega_0^2)} \tag{5.1}$$

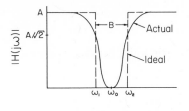

w (radians/second)   **Figure 5-1.** A band-reject amplitude response.

where, before the indicated substitution is made, the transfer function is
that of the nth order low-pass filter of Eq. (2.2). Evidently a $2n$th order band-
reject function arises from an nth order low-pass function, with a gain of G,
as in the low-pass case ($s = 0$ corresponds to $S = 0$).

The band-reject filter is a Butterworth or Chebyshev type when its
low-pass prototype is Butterworth or Chebyshev. A Butterworth band-reject
filter has monotonic amplitude response with a maximally flat passband. In
the Chebyshev band-reject filters, the passband has ripples, as in the low-
pass case, and the rate of attenuation at the cutoff points is greater than that
of the Butterworth. Again, except in the 3 dB ripple case, the frequencies ω_1
and ω_2 are the terminal frequencies of the stopband ripple channel in the
Chebyshev filter, and are the conventional cutoff points in the Butterworth
filter. In general, for Eq. (5.1), the center frequency is the geometric mean of
the cutoff frequencies. That is, $\omega_0 = \sqrt{\omega_1\omega_2}$, or in hertz, $f_0 = \sqrt{f_1f_2}$.
Amplitude and phase responses for fourth-order Butterworth and Chebyshev
designs are shown in Figs. 5-2 and 5-3, respectively.

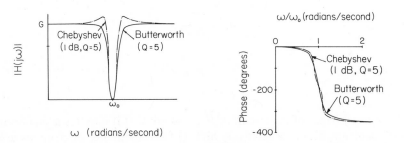

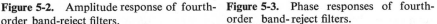

Figure 5-2. Amplitude response of fourth-   **Figure 5-3.** Phase responses of fourth-
order band-reject filters.   order band-reject filters.

Since first-order Butterworth and Chebyshev low-pass amplitude responses
have the same form except for scaling, a second-order band-reject function in
either case is given from Eq. (5.1) by taking $b_0 = 1$ and $n = 1$. The result is

$$\frac{V_2}{V_1} = \frac{G(s^2 + \omega_0^2)}{s^2 + Bs + \omega_0^2} \tag{5.2}$$

Higher-order responses, for which the Butterworth and Chebyshev cases are different, are given by Eq. (5.1) for $n = 2, 3, \ldots$.

5.2   VCVS Band-Reject Filters

A circuit credited to Inigo [41], which realizes the second-order band-reject function of Eq. (5.2) with a single op-amp, is shown in Fig. 5-4. The op-amp

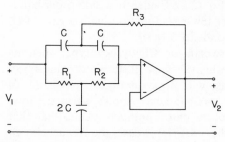

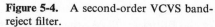

Figure 5-4.   A second-order VCVS band-reject filter.

with the unity feedback constitutes a VCVS with a gain of 1. Analysis of the circuit shows that Eq. (5.2) is realized if

$$\frac{1}{R_3} = \frac{1}{R_1} + \frac{1}{R_2} \tag{5.3}$$

and

$$B = \frac{2}{R_2 C}$$

$$\omega_0^2 = \frac{1}{R_1 R_2 C^2} \tag{5.4}$$

$$G = 1$$

Some advantages of the circuit of Fig. 5-4 are that it requires a small number of elements, attains a relatively high Q for the number of elements used, and has a noninverting gain. The center frequency f_0 can be adjusted, to some extent, by varying R_1, which leaves B (or Q) fixed. A disadvantage is that the gain is restricted to 1.

The procedure for rapid design of the band-reject filter of Fig. 5-4 is very similar to that of the filters described in the earlier chapters. The procedure is summarized in Sec. 5.5, where the general circuit is given followed by design Table 5-1, which may be used to read off practical circuit element values for a given center frequency f_0 and bandwidth B (or Q). (The gain is 1 in all cases.) The K parameter for use with the table may be found from

$$K = \frac{100}{f_0 C'} \tag{5.5}$$

where C' is the value of C in microfarads. Also, K may be read from the appropriate one of Figs. 5-8a, b, or c.

5.3 Multiple-Feedback Band-Reject Filters

Another circuit that realizes Eq. (5.2) is shown in Fig. 5-5 [7], which we refer to as a *multiple-feedback* band-reject filter. Analysis of the circuit shows that Eq. (5.2) is achieved if

$$R_3 R_4 = 2R_1 R_5 \tag{5.6}$$

with

$$B = \frac{2}{R_4 C}$$

$$\omega_0{}^2 = \frac{1}{R_4 C^2}\left(\frac{1}{R_1} + \frac{1}{R_2}\right) \tag{5.7}$$

$$G = -\frac{R_6}{R_3}$$

This circuit has the advantages of multiple-feedback structures described earlier in Sec. 2-8. It has a greater flexibility than the circuit of the previous section because the gain can be specified, and higher Q's are possible. The gain is inverting and can be set by varying R_6. The bandwidth B, and thus Q, can be varied to a limited degree, without affecting f_0, by changing R_4.

A summary of the techniques for obtaining a practical design of the band-reject filter of Fig. 5-5 is given in Sec. 5.6, together with Table 5-2, which is

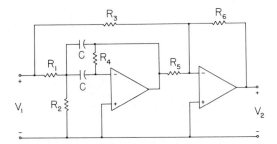

Figure 5-5. A second-order multiple-feedback band-reject filter.

the pertinent design table. An example is presented in [1] in detail, along with an actual response with a 60 Hz notch.

5.4 Biquad Band-Reject Filters

A biquad circuit that realizes the band-reject filter transfer function, credited to Fleischer and Tow [33], is shown in Fig. 5-6. Analysis of the circuit shows that Eq. (5.2) is achieved with

$$B = \frac{1}{R_2 C}$$

$$\omega_0{}^2 = \frac{1}{R_3 R_5 C^2} \tag{5.8}$$

$$G = -\frac{R_2}{R_1}$$

provided

$$R_2 R_6 = R_1 R_4 \tag{5.9}$$
$$R_4 R_7 = R_5 R_6$$

Although the biquad circuit requires more elements than the other circuits of this chapter, it is a very useful design because of its excellent tuning features. It can attain a much higher Q than the other circuits we have considered, and its relatively good stability makes cascading of two or more sections feasible for obtaining higher-order Butterworth and Chebyshev responses. In the case of a second-order biquad band-reject filter, the tuning

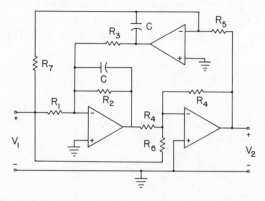

Figure 5-6. A second-order biquad band-reject filter.

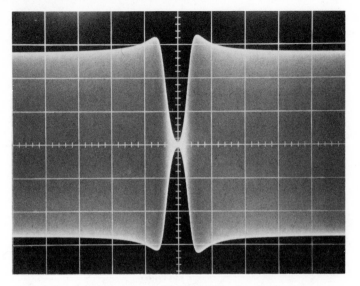

Figure 5-7. A fourth-order, 1 dB, Chebyshev band-reject response.

procedure is quite simple. To a limited degree, the gain can be adjusted by varying R_1 the bandwidth, and hence Q, can be adjusted by varying R_2, and the center frequency can be varied by changing R_3. In the cascading of two sections to obtain a fourth-order response, the passband characteristics can be varied by changing R_7.

The design procedure for the biquad band-reject filter is given in Sec. 5.7. Second-order designs can be performed using Table 5-3, and fourth-order cascaded designs are given in Tables 5-4 through 5-9.

As an example, suppose we want a fourth-order, 1 dB, Chebyshev band-reject filter with a gain of 100, $f_0 = 1000$ Hz, and $Q = 10$. Selecting $C = 0.01\mu$ F, we have from Eq. (5.5) a K parameter of 10. The filter will consist of two stages like Fig. 5-6, and the pertinent table is Table 5-7. Choosing each stage gain to be 10, we have from the table for stage 1, $R_1 = 33.32$, $R_2 = 333.21$, $R_3 = R_5 = 16.57$, $R_4 = 15.92$, $R_6 = 1.59$, and $R_7 = 1.53$ kΩ, and for stage 2, $R_1 = 30.72$, $R_2 = 307.23$, $R_3 = R_5 = 15.28$, $R_4 = 15.92$, $R_6 = 1.59$, and $R_7 = 1.66$ kΩ. These values are those of the table times the K parameter of 10. The resistances used were of 1% tolerance, with values as close as possible to the calculated values. The results were $Q = 11.8$, $f_0 = 1004$ Hz, $G = 100$, with a 1.1 dB ripple. The actual response is shown in Fig. 5-7.

5.5   Summary of VCVS Band-Reject Filter Design Procedure ($Q \leq 10$)

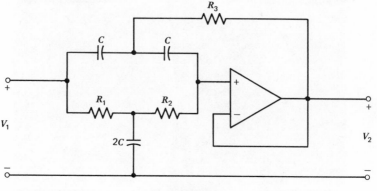

General circuit

Procedure

Given center frequency f_0 (hertz), gain = 1, and Q (or bandwidth $B = f_0/Q$), perform the following steps.

1.   Select a value of capacitance C and determine a K parameter from

$$K = \frac{100}{f_0 C'}$$

where C' is the value of C in microfarads. Alternatively, K may be found from Fig. 5-8a, b, or c.

2.   Find the resistance values from Table 5-1. The resistances in the table are given for $K = 1$, and hence their values must be multiplied by the K parameter of step 1 to yield the resistances of the circuit.

3.   Select standard resistance values that are as close as possible to those indicated by the table and construct the filter in accordance with the general circuit.

Comments and Suggestions

(a) For best performance, the input resistance of the op-amp should be at least 10 times $R_{eq} = R_1 + R_2$. For a specific op-amp, this condition can generally be satisfied by proper selection of C to obtain a suitable K parameter.

(b) Standard resistance values of 5% tolerance normally a yield acceptable results. For best performance, resistance values close to those indicated by the table should be used.

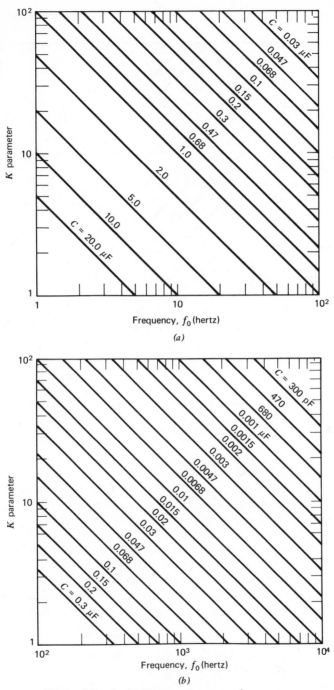

Figure 5-8. (*a, b*) *K* parameter versus frequency.

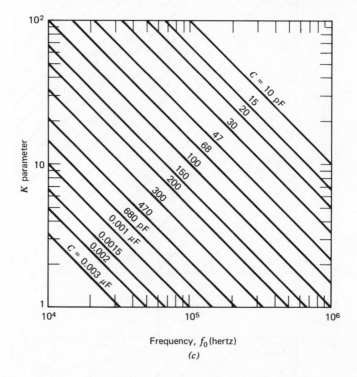

Figure 5-8.   (c) K parameter versus frequency.

In the case of capacitors, 5% tolerances should be used for best results. Since precision capacitors are relatively expensive, it may be desirable to use capacitors of higher tolerances, in which case trimming is generally required. In many cases, 10% capacitors are quite often satisfactory.

(c) There must be a dc return to ground at the filter input, the open-loop gain of the op-amp should be at least 50 times the gain of the filter at f_a, and the desired peak-to-peak voltage at f_a should not exceed $10^6/\pi f_a$ times the slew rate of the op-amp, where f_a is the highest frequency desired in the pass-band. Thus, for high values of f_a, externally compensated op-amps may be required.

(d) The center frequency f_0 can be adjusted, to some extent, by varying R_1, which, leaves B, and hence Q, fixed.

A discussion of the VCVS band-reject filters was given in Sec. 5.2.

Table 5-1. Second-Order VCVS
Band-Reject Filter Designs

Circuit Element Values[a]	
R_1	$0.796/Q$
R_2	$3.183Q$
R_3	$R_2/(4Q^2 + 1)$

[a] Resistances in kilohms
for a K parameter of 1,
gain is 1, quality factor is
Q.

5.6 Summary of Multiple-Feedback Band-Reject Filter Design Procedure ($Q \leq 25$)

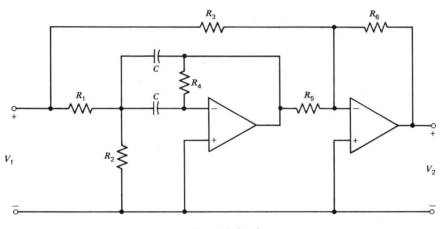

General circuit

Procedure

Given center frequency f_0 (hertz), gain G, and Q (or bandwidth $B = f_0/Q$), perform the following steps.

1. Select a value of capacitance C and determine a K parameter from

$$K = \frac{100}{f_0 C'}$$

where C' is the value of C in microfarads. Alternately, K may be found from Fig. 5-8a, b, or c.

2. Find the resistance values from Table 5-2. The resistances in the table are given for $K = 1$, and hence their values must be multiplied by the K-parameter of step 1 to yield the resistances of the circuit.

3. Select standard resistance values that are as close as possible to those indicated by the table and construct the filter in accordance with the general circuit.

Comments and Suggestions

The comments and suggestions for the VCVS band-reject filter given in Sec. 5.5 apply as follows:

(a) Paragraph (b) is directly applicable.

(b) Paragraph (c) applies except the dc return to ground is already satisfied by R_4.

In addition, the following applies:

(c) The inverting gain of the filter is R_6/R_3. Gain adjustments can be made by using a potentiometer in lieu of R_6. The bandwidth B can be tuned slightly by changing R_4, without affecting f_0.

The multiple-feedback back-reject filters were discussed in Sec. 5.3.

Table 5-2. Second-Order
Multiple-Feedback Band-Reject
Filter Designs

Circuit Element Values[a]	
R_1	$0.796Q$
R_2	$R_1/(Q^2 - 1)$
R_3	1.0
R_4	$4R_1$
R_5	2.0
R_6	G

[a] Resistances in kilohms for a K parameter of 1, gain is G (inverting), quality factor is Q.

5.7 Summary of Biquad Band-Reject Filter Design Procedure ($Q \leq 50$)

General circuit

Procedure

Given center frequency f_0 (hertz), gain G, order, and filter type (Butterworth or Chebyshev), perform the following steps for a second-order filter, or for each stage of a fourth-order cascaded filter.

1. Select a value of capacitance C and determine a K parameter from

$$K = \frac{100}{f_0 C'}$$

where C' is the value of C in microfarads. Alternatively, K may be found from Fig. 5-8a, b, or c. For fourth-order designs, it is better to use the equation since greater accuracy is required.

2. Find the resistance values from Table 5-3 in the second-order case and the appropriate one of Tables 5-4 through 5-9 in the fourth-order case. The resistances in the tables are given for $K = 1$, and hence their values must be multiplied by the K parameter of step 1 to yield the resistances of the circuit. The number G in the tables is the stage gain, and for fourth order,

the filter gain is the product of the stage gains. The stage gains, which are not necessarily equal, are chosen by the designer.

3.   Select standard resistance values that are as close as possible to those indicated by the table and construct the filter, or its stages, in accordance with the general circuit.

Comments and Suggestions

The comments and suggestions for the VCVS band-reject filter given in Sec. 5.5 apply as follows:

(a) Paragraph (b) applies in the second-order case. In the fourth-order case, resistors and capacitors of 2% and 5% tolerances, respectively, should be used.

(b) Paragraph (c) applies except that the dc return to ground is already satisfied by R_2 and R_3.

In addition, the following applies:

(c) The inverting stage gain is R_2/R_1. Varying R_1 affects the gain, varying R_2 affects Q, and varying R_3 changes f_0. In addition, in the fourth-order case, the passband response can be varied by changing R_7.

A specific example of a fourth-order design was given in Sec. 5.4.

Table 5-3.   Second-Order Biquad
Band-Reject Filter Designs

Circuit Element Values[a]	
R_1	$1.592Q/G$
R_2	$1.592Q$
R_3, R_4, R_5	1.592
R_6, R_7	$1.592/G$

[a] Resistances in kilohms for a
K parameter of 1, Q and G are
quality factor and gain.

Table 5-4. Fourth-Order Band-Reject Butterworth Cascaded Biquad Filter Designs

Circuit Element Values[a]

Q	1	2	3	4	5	6	8	10	50	Stage
R_1	3.467/G	5.468/G	7.654/G	9.876/G	12.110/G	14.350/G	18.839/G	23.333/G	113.341/G	
R_2	3.467	5.468	7.654	9.876	12.110	14.350	18.839	23.333	113.341	
R_3	2.296	1.903	1.792	1.739	1.708	1.688	1.664	1.649	1.603	1
R_4	1.592	1.592	1.592	1.592	1.592	1.592	1.592	1.592	1.592	
R_5	2.296	1.903	1.792	1.739	1.708	1.688	1.664	1.649	1.603	
R_6	1.592/G	1.592/G	1.592/G	1.592/G	1.592/G	1.592/G	1.592/G	1.592/G	1.592/G	
R_7	1.103/G	1.331/G	1.414/G	1.457/G	1.483/G	1.500/G	1.523/G	1.536/G	1.580/G	
R_1	1.666/G	3.826/G	6.041/G	8.272/G	10.511/G	12.753/G	17.244/G	21.739/G	111.750/G	
R_2	1.666	3.826	6.041	8.272	10.511	12.753	17.244	21.739	111.750	
R_3	1.103	1.331	1.414	1.457	1.483	1.500	1.523	1.536	1.580	2
R_4	1.592	1.592	1.592	1.592	1.592	1.592	1.592	1.592	1.592	
R_5	1.103	1.331	1.414	1.457	1.483	1.500	1.523	1.536	1.580	
R_6	1.592/G	1.592/G	1.592/G	1.592/G	1.592/G	1.592/G	1.592/G	1.592/G	1.592/G	
R_7	2.296/G	1.903/G	1.792/G	1.739/G	1.708/G	1.688/G	1.664/G	1.649/G	1.603/G	

[a] Resistances in kilohms for a K parameter of 1, G = stage gain, filter gain = product of stage gains.

Table 5-5. Fourth-Order Band-Reject Chebyshev Cascaded Biquad Filter Designs (0.1 dB)

Circuit Element Values[a]

Q	1	2	3	4	5	6	8	10	50	Stage
R_1	5.607/G	9.925/G	14.333/G	18.761/G	23.197/G	27.636/G	36.520/G	45.407/G	223.235/G	
R_2	5.607	9.925	14.333	18.761	23.197	27.636	36.520	45.407	223.235	
R_3	1.964	1.767	1.706	1.677	1.659	1.648	1.634	1.625	1.598	1
R_4	1.592	1.592	1.592	1.592	1.592	1.592	1.592	1.592	1.592	
R_5	1.964	1.767	1.706	1.677	1.659	1.648	1.634	1.625	1.598	
R_6	1.592/G	1.592/G	1.592/G	1.592/G	1.592/G	1.592/G	1.592/G	1.592/G	1.592/G	
R_7	1.290/G	1.434/G	1.485/G	1.511/G	1.527/G	1.537/G	1.551/G	1.559/G	1.585/G	
R_1	3.683/G	8.054/G	12.473/G	16.904/G	21.341/G	25.781/G	34.666/G	43.553/G	221.382/G	
R_2	3.683	8.054	12.473	16.904	21.341	25.781	34.666	43.553	221.382	
R_3	1.290	1.434	1.485	1.511	1.527	1.537	1.551	1.559	1.585	2
R_4	1.592	1.592	1.592	1.592	1.592	1.592	1.592	1.592	1.592	
R_5	1.290	1.434	1.485	1.511	1.527	1.537	1.551	1.559	1.585	
R_6	1.592/G	1.592/G	1.592/G	1.592/G	1.592/G	1.592/G	1.592/G	1.592/G	1.592/G	
R_7	1.963/G	1.767/G	1.706/G	1.677/G	1.659/G	1.648/G	1.634/G	1.625/G	1.598/G	

[a] Resistances in kilohms for a K parameter of 1, G = stage gain, filter gain = product of stage gains.

218

Table 5-6. Fourth-Order Band-Reject Chebyshev Cascaded Biquad Filter Designs (0.5 dB)

Circuit Element Values[a]

Q	1	2	3	4	5	6	8	10	50	Stage
R_1	4.990/G	8.103/G	11.412/G	14.761/G	18.126/G	21.497/G	28.251/G	35.012/G	170.396/G	
R_2	4.4990	8.103	11.412	14.761	18.126	21.497	28.251	35.012	170.396	
R_3	2.221	1.879	1.777	1.729	1.701	1.682	1.659	1.645	1.602	1
R_4	1.592	1.592	1.592	1.592	1.592	1.592	1.592	1.592	1.592	
R_5	2.221	1.879	1.777	1.729	1.701	1.682	1.659	1.645	1.602	
R_6	1.592/G	1.592/G	1.592/G	1.592/G	1.592/G	1.592/G	1.592/G	1.592/G	1.592/G	
R_7	1.140/G	1.348/G	1.425/G	1.465/G	1.490/G	1.506/G	1.527/G	1.540/G	1.581/G	
R_1	2.562/G	5.815/G	9.149/G	12.508/G	15.876/G	19.251/G	26.007/G	32.769/G	168.153/G	
R_2	2.562	5.815	9.149	12.508	15.876	19.251	26.007	32.769	168.153	
R_3	1.140	1.348	1.425	1.465	1.490	1.506	1.527	1.540	1.581	2
R_4	1.592	1.592	1.592	1.592	1.592	1.592	1.592	1.592	1.592	
R_5	1.140	1.348	1.425	1.465	1.490	1.506	1.527	1.540	1.581	
R_6	1.592/G	1.592/G	1.592/G	1.592/G	1.592/G	1.592/G	1.592/G	1.592/G	1.592/G	
R_7	2.221/G	1.879/G	1.777/G	1.729/G	1.701/G	1.682/G	1.659/G	1.645/G	1.602/G	

[a] Resistances in kilohms for a K parameter of 1, G = stage gain, filter gain = product of stage gains.

219

Table 5-7. Fourth-Order Band-Reject Chebyshev Cascaded Biquad Filter Designs (1 dB)

					Circuit Element Values[a]					
Q	1	2	3	4	5	6	8	10	50	Stage
R_1	5.198/G	7.996/G	11.082/G	14.227/G	17.394/G	20.572/G	26.942/G	33.321/G	161.156/G	
R_2	5.189	7.996	11.082	14.227	17.394	20.572	26.942	33.321	161.156	
R_3	2.388	1.950	1.822	1.726	1.726	1.703	1.674	1.657	1.605	1
R_4	1.592	1.592	1.592	1.592	1.592	1.592	1.592	1.592	1.592	
R_5	2.388	1.950	1.822	1.762	1.726	1.703	1.674	1.657	1.605	
R_6	1.592/G	1.592/G	1.592/G	1.592/G	1.592/G	1.592/G	1.592/G	1.592/G	1.592/G	
R_7	1.061/G	1.299/G	1.390/G	1.438/G	1.467/G	1.487/G	1.513/G	1.528/G	1.579/G	
R_1	2.308/G	5.327/G	8.453/G	11.613/G	14.787/G	17.968/G	24.341/G	30.723/G	158.560/G	
R_2	2.308	5.327	8.453	11.613	14.787	17.968	24.341	30.723	158.560	
R_3	1.061	1.299	1.390	1.438	1.467	1.487	1.513	1.528	1.579	2
R_4	1.592	1.592	1.592	1.592	1.592	1.592	1.592	1.592	1.599	
R_5	1.061	1.299	1.390	1.438	1.467	1.487	1.513	1.528	1.579	
R_6	1.592/G	1.592/G	1.592/G	1.592/G	1.592/G	1.592/G	1.592/G	1.592/G	1.592/G	
R_7	2.388/G	1.950/G	1.822/G	1.762/G	1.726/G	1.703/G	1.674/G	1.657/G	1.605/G	

[a] Resistances in kilohms for a K parameter of 1, G = stage gain, filter gain = product of stage gains.

Table 5-8. Fourth-Order Band-Reject Chebyshev Cascaded Biquad Filter Designs (2 dB)

Circuit Element Values[a]

Q	1	2	3	4	5	6	8	10	50	Stage
R_1	5.944/G	8.593/G	11.682/G	14.862/G	18.075/G	21.305/G	27.787/G	34.284/G	164.583/G	
R_2	5.944	8.593	11.682	14.862	17.075	21.305	27.787	34.284	164.583	
R_3	2.590	2.036	1.876	1.801	1.757	1.728	1.693	1.672	1.607	1
R_4	1.592	1.592	1.592	1.592	1.592	1.592	1.592	1.592	1.592	
R_5	2.590	2.036	1.876	1.801	1.757	1.728	1.693	1.672	1.607	
R_6	1.592/G	1.592/G	1.592/G	1.592/G	1.592/G	1.592/G	1.592/G	1.592/G	1.592/G	
R_7	0.978/G	1.244/G	1.350/G	1.407/G	1.442/G	1.466/G	1.496/G	1.515/G	1.576/G	
R_1	2.245/G	5.261/G	8.407/G	11.611/G	14.835/G	18.070/G	24.558/G	31.058/G	161.262/G	
R_2	2.245	5.251	8.407	11.611	14.835	18.070	24.558	31.058	161.362	
R_3	0.978	1.244	1.350	1.407	1.442	1.466	1.496	1.515	1.576	2
R_4	1.592	1.592	1.592	1.592	1.592	1.592	1.592	1.592	1.592	
R_5	0.978	1.244	1.350	1.407	1.442	1.466	1.496	1.515	1.576	
R_6	1.592/G	1.592/G	1.592/G	1.592/G	1.592/G	1.592/G	1.592/G	1.592/G	1.592/G	
R_7	2.590/G	2.036/G	1.876/G	1.801/G	1.757/G	1.728/G	1.693/G	1.672/G	1.607/G	

[a] Resistances in kilohms for a K parameter of 1, G = stage gain, filter gain = product of stage gains.

221

Table 5-9. Fourth-Order Band-Reject Chebyshev Cascaded Biquad Filter Designs (3 dB)

Q	1	2	3	4	5	6	8	10	50	Stage
				Circuit Element Values[a]						
R_1	6.833/G	9.523/G	12.791/G	16.180/G	19.614/G	23.069/G	30.009/G	36.970/G	176.654/G	
R_2	6.833	9.523	12.791	16.180	19.614	23.069	30.009	36.970	176.654	
R_3	2.715	2.091	1.910	1.825	1.776	1.744	1.705	1.681	1.609	1
R_4	1.592	1.592	1.592	1.592	1.592	1.592	1.592	1.592	1.592	
R_5	2.715	2.091	1.910	1.825	1.776	1.744	1.705	1.681	1.609	
R_6	1.592/G	1.592/G	1.592/G	1.592/G	1.592/G	1.592/G	1.592/G	1.592/G	1.592/G	
R_7	0.933/G	1.212/G	1.326/G	1.388/G	1.426/G	1.453/G	1.486/G	1.507/G	1.574/G	
R_1	2.347/G	5.520/G	8.880/G	12.302/G	15.751/G	19.214/G	26.163/G	33.127/G	172.818/G	
R_2	2.347	5.520	8.880	12.302	15.751	19.214	26.163	33.127	172.818	
R_3	0.933	1.212	1.326	1.388	1.426	1.453	1.486	1.507	1.574	2
R_4	1.592	1.592	1.592	1.592	1.592	1.592	1.592	1.592	1.592	
R_5	0.933	1.212	1.326	1.388	1.426	1.453	1.486	1.507	1.574	
R_6	1.592/G	1.592/G	1.592/G	1.592/G	1.592/G	1.592/G	1.592/G	1.592/G	1.592/G	
R_7	2.715/G	2.091/G	1.910/G	1.825/G	1.776/G	1.744/G	1.705/G	1.681/G	1.609/G	

[a] Resistances in kilohms for a K parameter of 1, G = stage gain, filter gain = product of stage gains.

6

Phase-Shift and Time-Delay Filters

6.1 All-Pass Filters

An *all-pass*, or *phase-shifting*, filter is one that passes signals of all frequencies equally well while changing or shifting their phase by some prescribed amount. In other words, the amplitude response is ideally constant and the phase response varies with frequency. A typical phase response $\phi(\omega)$ of an all-pass filter is shown in Fig. 6-1, where it may be seen that the phase shift varies with ω. Since shifting a frequency by some negative amount is equivalent to delaying that component by some positive time as it passes through the filter, the all-pass filter may also be thought of as a time-delay circuit.

The transfer functions we consider are ratios V_2/V_1 of output to input voltages. Thus, at ω_0 (or in hertz, f_0), if the phase is a negative number, say, $\phi(\omega_0) = -\phi_0$ degrees ($\phi_0 > 0$), then at ω_0 the phase of the input V_1 is greater than that of the output V_2 by ϕ_0 degrees. If the two waveforms are sinusoids and are viewed simultaneously, the input wave reaches peaks or dips ϕ_0 degrees before the output wave reaches its peaks or dips. Therefore, the input signal is leading the output signal by ϕ_0 degrees. Also, the difference in time in seconds between a peak or dip of the input wave and the immediately succeeding peak or dip of the output wave, when the amplitudes are plotted versus time, is the time-delay. Evidently a phase shift of $-\phi_0$ is equivalent to a phase shift of $360° - \phi_0$. For example, if the input wave leads the output wave by $270°$ ($\phi = -\phi_0 = -270°$), then it may also be said that the input leads by $-90°$, or equivalently, the output leads by $+90°$. Figure 6-2 shows

223

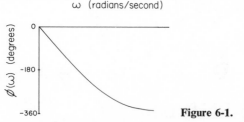

Figure 6-1. A typical phase response.

an output wave lagging an input wave (or the input is leading the output) by an amount ϕ_0. If the horizontal axis were in seconds, the difference between the two successive peaks would be the time-delay in seconds.

We may note here that if the output wave were simply the input wave delayed in time by an amount T_d (and possibly amplified), then T_d is given by

$$T_d = -\frac{d}{d\omega}\,\phi(\omega) \tag{6.1}$$

which is the same as the time-delay τ defined earlier in Sec. 2.2. We shall take Eq. (6.1) as our definition of time-delay later in Sec. 6.4, where it is important in the case of the Bessel filter.

A second-order all-pass filter transfer function is given by

$$H(s) = \frac{V_2}{V_1} = \frac{G(s^2 - as + b)}{s^2 + as + b} \tag{6.2}$$

where a and b are constants. This may be seen by computing the amplitude, given by $|H(j\omega)| = G$, which is also defined to be the gain of the filter. The phase response is given by

$$\phi(\omega) = -2\arctan\left(\frac{a\omega}{b - \omega^2}\right) \tag{6.3}$$

and may also be given in terms of $f = \omega/2\pi$ Hz. Thus, a and b may be varied to give a prescribed phase shift at a given frequency, while the amplitude remains constant.

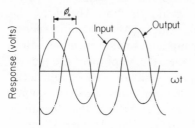

Figure 6-2. Two waves with different phases.

6.2 Second-Order Single Op-Amp All-Pass Filters

An all-pass network that achieves Eq. (6.2) with a single op-amp is shown in Fig. 6-3 [42]. Analysis of the circuit yields the function of Eq. (6.2) with

$$a = \frac{2}{R_2 C}$$

$$b = \frac{1}{R_1 R_2 C^2} \tag{6.4}$$

$$G = \frac{R_4}{R_3 + R_4}$$

provided that

$$R_2 R_3 = 4 R_1 R_4 \tag{6.5}$$

The designer of an all-pass filter using Fig. 6-3 may specify the phase shift desired at a given frequency and obtain practical element values as described in the summary in Sec. 6.7. As in the case of the filters in the earlier chapters, a K parameter is first found from

$$K = \frac{100}{f_0 C'} \tag{6.6}$$

where C' is the value of C in microfarads in Fig. 6-3, and f_0 is the frequency at which a specific shift ϕ_0 is desired. Alternatively, K may be read off Fig. 6-8a, b, or c, depending on f_0.

The primary advantage of the circuit of Fig. 6-3 is that it requires a small number of elements to produce an all-pass response. The phase-shift can be varied, to a limited degree, by changing R_2 and the gain varied by changing R_4. The gain, however, is restricted to values less than one. The circuit resistances are given in the appropriate one of Tables 6-1 through 6-12 for a K parameter of 1, and hence they must be multiplied by K obtained from Eq. (6.6). The tables are designed to give a gain of $\frac{1}{2}$ in every case, for phase-shifts of $\pm 5°$ to $\pm 175°$ in $5°$ increments.

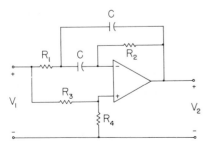

Figure 6-3. A second-order all-pass filter.

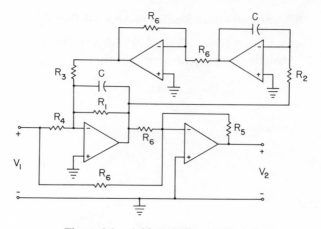

Figure 6-4. A biquad all-pass circuit.

6.3 Biquad All-Pass Filters

A biquad circuit [27] that realizes Eq. (6.2) is shown in Fig. 6-4. Analysis yields

$$a = \frac{1}{R_1 C}$$

$$b = \frac{1}{R_2 R_3 C^2} \qquad (6.7)$$

$$G = -\frac{R_5}{R_6}$$

provided

$$R_1 = 2R_4 \qquad (6.8)$$

The biquad circuit requires more elements than the circuit of the previous section, but it is more stable and very easy to tune. The gain may be adjusted by varying R_5. Changing R_1 affects the parameter a and changing R_3 affects the parameter b which, in turn, varies the phase-shift.

The design procedure for rapid construction of the biquad all-pass filter is given in Sec. 6.8, and the pertinent design tables are Tables 6-13 through 6.24. The tables are designed to give an inverting gain of 1 in every case for phase-shifts of $\pm 5°$ to $\pm 175°$ in $5°$ increments.

6.4 Constant-Time-Delay or Bessel Filters

A filter for which the phase response shown in Fig. 6-1 is a straight line sloping downward to the right is a *linear-phase*, or *constant-time-delay* filter.

This is evident from Eq. (6.1) since, in this case, the phase-shift is proportional to the frequency, given by $\phi(\omega) = -k\omega$, and thus $T_d = -d(-k\omega)/d\omega = k$ is constant for all frequencies.

The all-pole filter (see Sec. 2.1) that best approximates a constant-time-delay circuit is the Bessel filter [43]. In the second-order case it has a transfer function

$$\frac{V_2}{V_1} = \frac{3\omega_0^2 G}{s^2 + 3\omega_0 s + 3\omega_0^2} \tag{6.9}$$

In the third-order case, the transfer function is

$$\frac{V_2}{V_1} = \frac{15\omega_0^3 G}{s^3 + 6\omega_0 s^2 + 15\omega_0^2 s + 15\omega_0^3} \tag{6.10}$$

and in the fourth-order case, the transfer function is

$$\frac{V_2}{V_1} = \frac{105\omega_0^4 G}{s^4 + 10\omega_0 s^3 + 45\omega_0^2 s^2 + 105\omega_0^3 s + 105\omega_0^4} \tag{6.11}$$

In every case the gain is G, the phase is approximately linear, and the time-delay is approximately $1/\omega_0$ for frequencies from 0 to $f_0 = \omega_0/2\pi$ Hz. The approximation improves as the order increases. For example, in every case the time-delay at $\omega = 0$ is $T_d = 1/\omega_0$ seconds. At $\omega = \omega_0$, for second order, $T_d = 12/13\omega_0$, for third order, $T_d = 276/277\omega_0$, and for fourth order, $T_d = 12{,}745/12{,}746\omega_0$. Equations (6.9), (6.10), and (6.11) are special cases obtained from a general continued fraction expansion, and may be readily extended to higher orders [22].

To illustrate the quality of the Bessel filter and the improvement with order, we have shown in Fig. 6-5 the phase responses of a second- and fourth-order Bessel filter, and for comparative purposes, that of a fourth-order Butterworth filter. Evidently, both the Bessel responses are far superior to that of

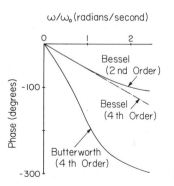

Figure 6-5. Bessel and Butterworth phase responses.

the Butterworth, and a fourth-order Bessel response is superior to a second order.

The amplitude response of the Bessel filter resembles somewhat that of an all-pass filter except that it drops slowly and monotonically from its maximum value, which occurs at zero frequency. This is to be expected, however, since the transfer function in every case resembles that of a low-pass filter.

6.5 Bessel Filter Realizations

Since the transfer function of the Bessel filter is identical in form to that of a low-pass filter, it may be realized by any of the circuits described in Chapter 2. The design procedure, which is summarized in Sec. 6.9, is identical to those of Chapter 2 except that instead of the cutoff frequency f_c, we are interested in the frequency f_0 associated with the time-delay. At zero frequency, the time-delay is given by $T_d = 1/\omega_0$, and hence if T_d is in seconds we have

$$f_0 = \frac{0.15915}{T_d} \text{ Hz} \qquad (6.12)$$

This is true for all orders and T_d is approximately constant for frequencies from 0 to f_0 Hz, an approximation that improves as the order increases.

We first need to find f_0 which, upon selection of the capacitance C in the circuit, is used to obtain the K parameter in Eq. (6.6). We may find f_0 from Eq. (6.12) for the value of T_d desired, or f_0 may be read from the appropriate one of Figs. 6-9a, b, or c. The rest of the design procedure is identical to those of Chapter 2.

We have given tables for the Bessel filter elements for the VCVS design of Sec. 2.5 (Tables 6-25 and 6-26) and the infinite-gain MFB design of Sec. 2.6 (Tables 6-27 and 6-28) for the second- and fourth-order cases. In addition, we have provided third- and fourth-order designs in Tables 6-29 and 6-30 which utilize the multiple-feedback circuits of Sec. 2.8.

As an example, let us design a fourth-order multiple-feedback Bessel filter with a constant-time-delay T_d of 159.15 microseconds (μsec) and a gain of 4. By Eq. (6.12) we have $f_0 = 1000$ Hz, which may also be found from Fig. 6-9b. Selecting $C = 0.01 \mu$F, we have from Eq. (6.6) a K parameter of 10. The general circuit is that of Fig. 2-10, and the applicable table is Table 6-30. The resistance values are multiplied by the K parameter of 10, and are given by $R_1 = 1.35$, $R_2 = 15.83$, $R_3 = 2.07$, $R_4 = 6.92$, $R_5 = 11.98$, and $R_6 = 35.95$ kΩ. The actual results were $T_d = 160 \mu$sec at 1000 Hz and at 2000 Hz, and a gain of 4. The gain is taken as in the low-pass case, to be the amplitude function at zero frequency. At 1000 Hz, this function was 3.8, and at 2000 Hz, it was 2.9. Both the amplitude response and the input and output waveforms are shown in Fig. 6-6. The amplitude response was swept from 0 to 2000 Hz and the waveform horizontal scale was 200 μsec/div.

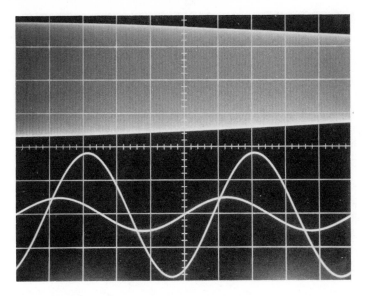

Figure 6-6. Amplitude response and waveforms of a fourth-order Bessel filter.

6.6 All-Pass Constant-Time-Delay Filters

Let the transfer function of the Bessel filter be given by

$$H(s) = \frac{K}{P(s)} \tag{6.13}$$

where K is a constant and $P(s)$ is the denominator polynomial, given for second-, third-, and fourth-order cases in Eqs. (6.9), (6.10), and (6.11). Then the function

$$\frac{V_2}{V_1} = G \frac{P\left(\dfrac{-s}{2}\right)}{P\left(\dfrac{s}{2}\right)} \tag{6.14}$$

is a transfer function of a circuit that is both an all-pass and a constant-time-delay filter [44].

An advantage of a constant-time-delay filter that is also all-pass is that the amplitude response does not decay with increasing frequency, as in the case of the Bessel filter. In addition, it is shown in [44] that the time-delay is constant over twice the frequency range of the Bessel filter.

For the second-order Bessel filter case, Eq. (6.14) becomes

$$\frac{V_2}{V_1} = \frac{G(s^2 - 6\omega_0 s + 12\omega_0^2)}{s^2 + 6\omega_0 s + 12\omega_0^2} \tag{6.15}$$

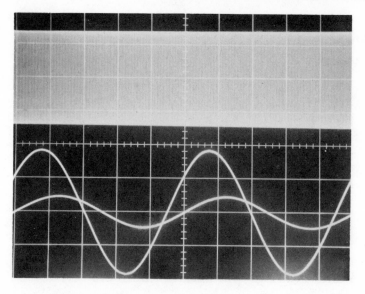

Figure 6-7. Amplitude response and waveforms of an all-pass, constant-time-delay filter.

and thus can be realized with any of the three all-pass circuits considered earlier in this chapter. In this section, we shall consider a practical design using the single op-amp configuration of Sec. 6.2. Equations (6.4) and (6.5) hold with $a = 6\omega_0$ and $b = 12\omega_0^2$. The general circuit and design procedure are given in Sec. 6.10, and the pertinent design table is Table 6-31, for a gain of 0.25.

As an example, suppose we want a second-order all-pass constant-time-delay filter with $T_d = 100\,\mu\text{sec}$. From Eq. (6.12) we have $f_0 = 1592\,\text{Hz}$. Selecting $C = 0.01\,\mu\text{F}$ we have from Eq. (6.6) a K parameter of 6.28. Multiplying the resistances of Table 6-31 by 6.28 we have $R_1 = 2.5$, $R_2 = 3.33$, $R_3 = 13.33$, and $R_4 = 4.44\,\text{k}\Omega$. The circuit was constructed using resistances of 2.5, 3.32, 13, and 4.7 kΩ, resulting in $T_d = 100\,\mu\text{sec}$ at 1000, 1592, and 2000 Hz, and a constant gain of 4. Figure 6-7 shows the all-pass amplitude response and the input and output waves at 1000 Hz.

6.7 Summary of Single Op-Amp All-Pass Filter Design Procedure

Procedure

Given f_0 (hertz) and the phase-shift ϕ desired at f_0, perform the following steps.

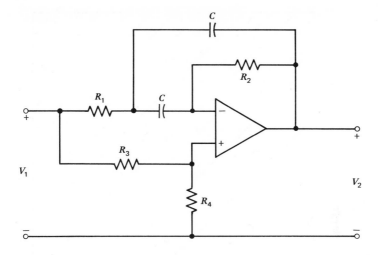

General circuit

1. Select a value of capacitance C and determine a K parameter from

$$K = \frac{100}{f_0 C'}$$

where C' is the value of C in microfarads. Alternatively, K may be found from Fig. 6-8 a, b, or c.

2. Find the resistance values from the appropriate one of Tables 6-1 through 6-12, depending on ϕ. The resistances in the tables are given for $K = 1$, and hence their values must be multiplied by the K parameter of step 1 to yield the resistances of the circuit.

3. Select standard resistance values that are as close as possible to those indicated by the table and construct the filter in accordance with the general circuit.

Comments and Suggestions

(a) Standard resistance values of 5% tolerance normally yield acceptable results. For best performance, resistance values close to those indicated by the tables should be used.

In the case of capacitors, 5% tolerances should be used for best results. Since precision capacitors are relatively expensive, it may be desirable to use capacitors of higher tolerances, in which case trimming is generally required. In many cases 10% capacitors are quite often satisfactory.

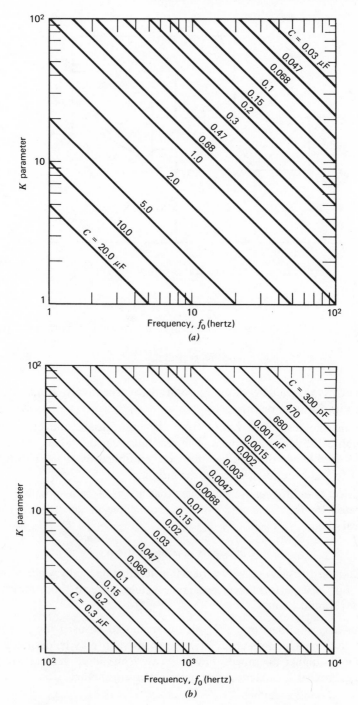

(a)

(b)

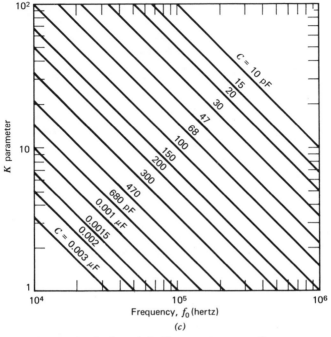

Figure 6-8. (*a, b,* and *c*). *K* parameter versus frequency.

(b) The open-loop gain of the op-amp should be at least 50 times the gain of the filter at f_0, and the desired peak-to-peak voltage at f_0 should not exceed $10^6/\pi f_0$ times the slew rate of the op-amp. Thus, for high values of f_0, externally compensated op-amps may be required.

(c) The phase-shift and gain can be tuned by varying R_2 and R_4, respectively.

A discussion of the single op-amp all-pass filters was given in Sec. 6.2.

Table 6-1. All-Pass (Phase-Shift) Single Op-Amp Filter Designs (5° to 30°)

	Circuit Element Values[a]					
Shift	5°	10	15°	20°	25°	30°
R_1	18.261	9.165	6.148	4.649	3.758	3.170
R_2	73.044	36.659	24.590	18.597	15.032	12.679
R_3, R_4	146.087	73.319	49.180	37.194	30.064	25.357

[a] Resistances in kilohms for a *K* parameter of 1, gain is 0.5.

Table 6-2. All-Pass (Phase-Shift) Single Op-Amp Filter Designs
(35° to 60°)

Shift	Circuit Element Values[a]					
	35°	40°	45°	50°	55°	60°
R_1	2.754	2.445	2.208	2.020	1.868	1.742
R_2	11.015	9.781	8.832	8.080	7.471	6.967
R_3, R_4	22.031	19.563	17.664	16.160	14.942	13.935

[a] Resistances in kilohms for a K parameter of 1, gain is 0.5.

Table 6-3. All-Pass (Phase-Shift) Single Op-Amp Filter Designs
(65° to 90°)

Shift	Circuit Element Values[a]					
	65°	70°	75°	80°	85°	90°
R_1	1.636	1.546	1.468	1.401	1.341	1.288
R_2	6.545	6.184	5.873	5.602	5.363	5.150
R_3, R_4	13.089	12.369	11.747	11.204	10.726	10.301

[a] Resistances in kilohms for a K parameter of 1, gain is 0.5.

Table 6-4. All-Phase (Phase-Shift) Single Op-Amp Filter Designs
(95° to 120°)

Shift	Circuit Element Values[a]					
	95°	100°	105°	110°	115°	120°
R_1	1.240	1.197	1.158	1.122	1.089	1.058
R_2	4.960	4.787	4.631	4.487	4.355	4.232
R_3, R_4	9.919	9.575	9.261	8.974	8.709	8.464

[a] Resistances in kilohms for a K parameter of 1, gain is 0.5.

Table 6-5. All-Pass (Phase-Shift) Single Op-Amp Filter Designs

	Circuit Element Values[a]					
Shift	125°	130°	135°	140°	145°	150°
R_1	1.029	1.003	0.977	0.954	0.931	0.909
R_2	4.118	4.011	3.910	3.815	3.724	3.638
R_3, R_4	8.235	8.021	7.820	7.629	7.448	7.276

[a] Resistances in kilohms for a K parameter of 1, gain is 0.5.

Table 6-6. All-Pass (Phase-Shift) Single Op-Amp Filter Designs (155° to 175°)

	Circuit Element Values[a]				
Shift	155°	160°	165°	170°	175°
R_1	0.889	0.869	0.850	0.831	0.813
R_2	3.555	3.476	3.400	3.325	3.253
R_3, R_4	7.111	6.952	6.799	6.651	6.507

[a] Resistances in kilohms for a K parameter of 1, gain is 0.5.

Table 6-7. All-Pass (Phase-Shift) Single Op-Amp Filter Designs ($-5°$ to $-30°$)

	Circuit Element Values[a]					
Shift	$-5°$	$-10°$	$-15°$	$-20°$	$-25°$	$-30°$
R_1	0.035	0.069	0.103	0.136	0.169	0.200
R_2	0.139	0.276	0.412	0.545	0.674	0.799
R_3, R_4	0.277	0.553	0.824	1.090	1.348	1.598

[a] Resistances in kilohms for a K parameter of 1, gain is 0.5.

Table 6-8.   All-Pass (Phase-Shift) Single Op-Amp Filter Designs
($-35°$ to $-60°$)

Shift	Circuit Element Values[a]					
	$-35°$	$-40°$	$-45°$	$-50°$	$-55°$	$-60°$
R_1	0.230	0.259	0.287	0.313	0.339	0.364
R_2	0.920	1.036	1.147	1.254	1.356	1.454
R_3, R_4	1.840	2.072	2.294	2.508	2.712	2.908

[a] Resistances in kilohms for a K parameter of 1, gain is 0.5.

Table 6-9.   All-Pass (Phase-Shift) Single Op-Amp Filter Designs
($-65°$ to $-90°$)

Shift	Circuit Element Values[a]					
	$-65°$	$-70°$	$-75°$	$-80°$	$-85°$	$-90°$
R_1	0.387	0.410	0.431	0.452	0.472	0.492
R_2	1.548	1.638	1.725	1.809	1.889	1.967
R_3, R_4	3.096	3.277	3.450	3.617	3.779	3.935

[a] Resistances in kilohms for a K parameter of 1, gain is 0.5.

Table 6-10.   All-Pass (Phase-Shift) Single Op-Amp Filter Designs
($-95°$ to $120°$)

Shift	Circuit Element Values[a]					
	$-95°$	$-100°$	$-105°$	$-110°$	$-115°$	$-120°$
R_1	0.511	0.529	0.547	0.565	0.582	0.599
R_2	2.043	2.116	2.188	2.258	2.327	2.394
R_3, R_4	4.086	4.233	4.376	4.516	4.654	4.788

[a] Resistances in kilohms for a K parameter of 1, gain is 0.5.

Table 6-11. All-Pass (Phase-Shift) Single Op-Amp Filter Designs
($-125°$ to $-150°$)

Shift	Circuit Element Values[a]					
	$-125°$	$-130°$	$-135°$	$-140°$	$-145°$	$-150°$
R_1	0.615	0.632	0.648	0.664	0.680	0.696
R_2	2.461	2.526	2.591	2.656	2.721	2.785
R_3, R_4	4.921	5.053	5.183	5.312	5.441	5.570

[a] Resistances in kilohms for a K parameter of 1, gain is 0.5.

Table 6-12. All-Pass (Phase-Shift) Single Op-Amp Filter
Designs ($-155°$ to $-175°$)

Shift	Circuit Element Values[a]				
	$-155°$	$-160°$	$-165°$	$-170°$	$-175°$
R_1	0.712	0.729	0.745	0.762	0.779
R_2	2.850	2.915	2.980	3.047	3.114
R_3, R_4	5.700	5.830	5.961	6.094	6.229

[a] Resistances in kilohms for a K parameter of 1, gain is 0.5.

6.8 Summary of Biquad All-Pass Filter Design Procedure

Procedure

Given f_0 (hertz) and the phase shift ϕ desired at f_0, perform the following steps.

1. Select a value of capacitance C and determine a K parameter from

$$K = \frac{100}{f_0 C'}$$

where C' is the value of C in microfarads. Alternately, K may be found from Fig. 6-8a, b, or c.

2. Find the resistance values from the appropriate one of Tables 6-13 through 6-24. The resistances in the tables are given for $K = 1$ and hence their values must be multiplied by the K parameter of step 1 to yield the resistances of the circuit.

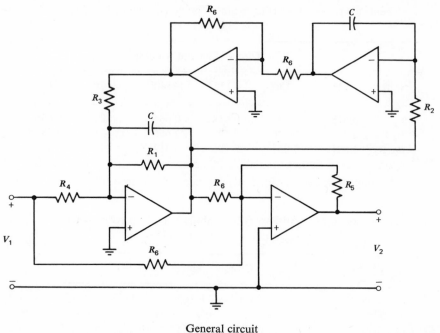

General circuit

3. Select standard resistance values that are as close as possible to those indicated by the table and construct the filter in accordance with the general circuit.

Comments and Suggestions

(a) The comments and suggestions for the single op-amp all-pass filter given in paragraphs (a) and (b) of Sec. 6.7 apply directly.

(b) In addition, the filter response is readily tuned by varying R_1, R_3, and R_5. The phase-shift can be tuned by varying R_1 and R_3. The gain can be tuned by changing R_5.

A discussion of the biquad all-pass filters is given in Sec. 6.3.

Table 6-13.   All-Pass (Phase-Shift) Biquad Filter Designs
(5° to 30°)

Shift	Circuit Element Values[a]					
	5°	10°	15°	20°	25°	30°
R_1, R_2, R_3	36.522	18.330	12.295	9.299	7.516	6.339
R_4	18.261	9.165	6.148	4.649	3.758	3.170
R_5, R_6	1.592	1.592	1.592	1.592	1.592	1.592

[a] Resistances in kilohms for a K parameter of 1, gain is 1 (inverting).

Table 6-14.   All-Pass (Phase-Shift) Biquad Filter Designs
(35° to 60°)

Shift	Circuit Element Values[a]					
	35°	40°	45°	50°	55°	60°
R_1, R_2, R_3	5.508	4.891	4.416	4.040	3.735	3.484
R_4	2.754	2.445	2.208	2.020	1.868	1.742
R_5, R_6	1.592	1.592	1.592	1.592	1.592	1.592

[a] Resistances in kilohms for a K parameter of 1, gain is 1 (inverting).

Table 6-15.   All-Pass (Phase Shift) Biquad Filter Designs
(65° to 90°)

Shift	Circuit Element Values[a]					
	65°	70°	75°	80°	85°	90°
R_1, R_2, R_3	3.272	3.092	2.937	2.801	2.682	2.575
R_4	1.636	1.546	1.468	1.401	1.341	1.288
R_5, R_6	1.592	1.592	1.592	1.592	1.592	1.592

[a] Resistances in kilohms for a K parameter of 1, gain is 1 (inverting).

Table 6-16. All-Pass (Phase-Shift) Biquad Filter Designs
(95° to 120°)

Shift	Circuit Element Values[a]					
	95°	100°	105°	110°	115°	120°
R_1, R_2, R_3	2.480	2.394	2.315	2.243	2.177	2.116
R_4	1.240	1.197	1.158	1.122	1.089	1.058
R_5, R_6	1.592	1.592	1.592	1.592	1.592	1.592

[a] Resistances in kilohms for a K parameter of 1, gain is 1 (inverting).

Table 6-17. All-Pass (Phase-Shift) Biquad Filter Designs
(125° to 150°)

Shift	Circuit Element Values[a]					
	125°	130°	135°	140°	145°	150°
R_1, R_2, R_3	2.059	2.005	1.955	1.907	1.862	1.819
R_4	1.029	1.003	0.977	0.954	0.931	0.909
R_5, R_6	1.592	1.592	1.592	1.592	1.592	1.592

[a] Resistances in kilohms for a K parameter of 1, gain is 1 (inverting).

Table 6-18. All-Pass (Phase-Shift) Biquad Filter Designs
(155° to 175°)

Shift	Circuit Element Values[a]				
	155°	160°	165°	170°	175°
R_1, R_2, R_3	1.778	1.738	1.700	1.663	1.627
R_4	0.889	0.869	0.850	0.831	0.813
R_5, R_6	1.592	1.592	1.592	1.592	1.592

[a] Resistances in kilohms for a K parameter of 1, gain is 1 (inverting).

Table 6-19.   All-Pass (Phase-Shift) Biquad Filter Designs
($-5°$ to $-30°$)

Shift	Circuit Element Values[a]					
	$-5°$	$-10°$	$-15°$	$-20°$	$-25°$	$-30°$
R_1, R_2, R_3	0.069	0.138	0.206	0.272	0.337	0.400
R_4	0.035	0.069	0.103	0.136	0.169	0.200
R_5, R_6	1.592	1.592	1.592	1.592	1.592	1.592

[a] Resistances in kilohms for a K parameter of 1, gain is 1 (inverting).

Table 6-20.   All-Pass (Phase-Shift) Biquad Filter Designs
($-35°$ to $-60°$)

Shift	Circuit Element Values[a]					
	$-35°$	$-40°$	$-45°$	$-50°$	$-55°$	$-60°$
R_1, R_2, R_3	0.460	0.518	0.574	0.627	0.678	0.727
R_4	0.230	0.259	0.287	0.313	0.339	0.364
R_5, R_6	1.592	1.592	1.592	1.592	1.592	1.592

[a] Resistances in kilohms for a K parameter of 1, gain is 1 (inverting).

Table 6-21.   All-Pass (Phase-Shift) Biquad Filter Designs
($-65°$ to $-90°$)

Shift	Circuit Element Values[a]					
	$-65°$	$-70°$	$-75°$	$-80°$	$-85°$	$-90°$
R_1, R_2, R_3	0.774	0.819	0.863	0.904	0.945	0.984
R_4	0.387	0.410	0.431	0.452	0.472	0.492
R_5, R_6	1.592	1.592	1.592	1.592	1.592	1.592

[a] Resistances in kilohms for a K parameter of 1, gain is 1 (inverting).

Table 6-22. All-Pass (Phase-Shift) Biquad Filter Designs (−95° to −120°)

Shift	Circuit Element Values[a]					
	−95°	−100°	−105°	−110°	−115°	−120°
R_1, R_2, R_3	1.021	1.058	1.094	1.129	1.163	1.197
R_4	0.511	0.529	0.547	0.565	0.582	0.599
R_5, R_6	1.592	1.592	1.592	1.592	1.592	1.592

[a] Resistances in kilohms for a K parameter of 1, gain is 1 (inverting).

Table 6-23. All-Pass (Phase-Shift) Biquad Filter Designs (−125° to −150°)

Shift	Circuit Element Values[a]					
	−125°	−130°	−135°	−140°	−145°	−150°
R_1, R_2, R_3	1.230	1.263	1.296	1.328	1.360	1.393
R_4	0.615	0.632	0.648	0.664	0.680	0.696
R_5, R_6	1.592	1.592	1.592	1.592	1.592	1.592

[a] Resistances in kilohms for a K parameter of 1, gain in 1 (inverting).

Table 6-24. All-Pass (Phase-Shift) Biquad Filter Designs
(−155° to 175°)

Shift	Circuit Element Values[a]				
	−155°	−160°	−165°	−170°	−175°
R_1, R_2, R_3	1.425	1.457	1.490	1.523	1.557
R_4	0.712	0.729	0.745	0.762	0.779
R_5, R_6	1.592	1.592	1.592	1.592	1.592

[a] Resistances in kilohms for a K parameter of 1, gain is 1 (inverting).

6.9   Summary of Bessel (Constant-Time-Delay) Filter Design Procedure

Procedure

Given time-delay T_d, order (second, third, or fourth), and gain, perform the following steps.

1.   Find f_0 in hertz for T_d in seconds from

$$f_0 = \frac{0.15915}{T_d}$$

or from the appropriate one of Fig. 6-9a, b, or c. (Note: Alternately, one could specify f_0 and find T_d by this procedure.)

2.   Select a value of capacitance C and determine a K parameter from

$$K = \frac{100}{f_0 C'}$$

where C' is the value of C in microfarads. Alternately, K may be found from Fig. 6-8a, b, or c, but for the fourth-order case it is better to use the equation since higher accuracy is required.

3.   Select the circuit desired to implement the filter from one of the low-pass filters, VCVS, infinite-gain MFB, or multiple-feedback (see Sec. 6.5). Using the value of K obtained in step 2, find the resistance values from the appropriate table, as follows:

(a) If the VCVS circuit is chosen, the general circuit and comments and suggestions are given in Sec. 2.10, and the appropriate tables are Tables 6-25 and 6-26 for second and fourth orders, respectively.

(b) If the infinite-gain MFB circuit is chosen, the general circuit and comments and suggestions are given in Sec. 2.11, and the appropriate tables are Tables 6-27 and 6-28 for second and fourth orders, respectively.

(c) If the multiple feedback circuit is chosen, the general circuits for third and fourth orders are given in Figs. 2-9 and 2-10, respectively. Comments and suggestions are given in Sec. 2.13. The appropriate tables are Tables 6-29 and 6-30 for third and fourth orders, respectively.

Comments and Suggestions

The comments and suggestions indicated in (a), (b), and (c) above which concern f_c in Chapter 2 now apply to f_0.

A specific example of a fourth-order multiple-feedback Bessel filter was given in Sec. 6.5.

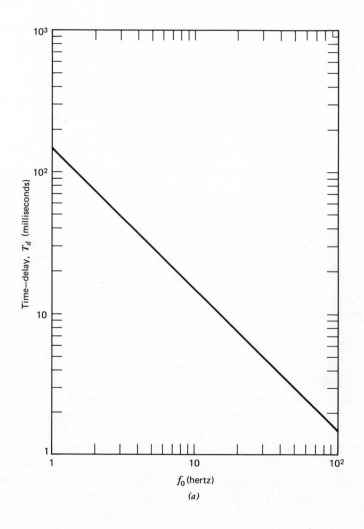

(a)

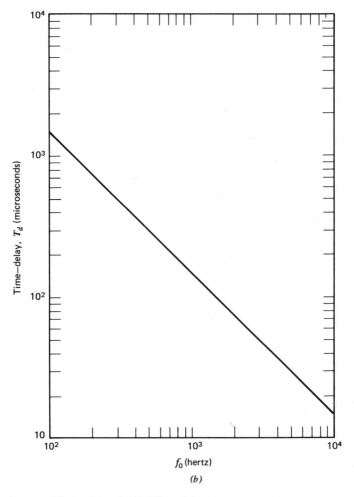

Figure 6-9. (a, b). Time-delay versus frequency.

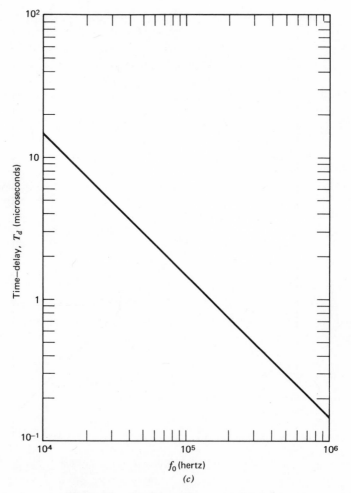

Figure 6-9. (*c*). Time-delay versus frequency.

Table 6-25. Second-Order Bessel (Constant-Time-Delay)
VCVS Filter Designs

	Circuit Element Values[a]					
Gain	1	2	4	6	8	10
R_1	0.673	0.531	0.420	0.328	0.282	0.252
R_2	2.510	1.592	1.006	1.288	1.500	1.677
R_3	Open	4.244	1.901	1.939	2.036	2.143
R_4	0	4.244	5.702	9.693	14.249	19.287
C_1	0.5C	C	2C	2C	2C	2C

[a] Resistances in kilohms for a K parameter of 1.

Table 6-26. Fourth-Order Bessel (Constant-Time-Delay) Cascaded VCVS
Filter Designs

	Circuit Element Values[a]						
Gain	1	2	6	10	36	100	Stage
R_1	0.328	0.275	0.225	0.179	0.179	0.140	
R_2	1.689	1.009	1.233	1.546	0.773	0.993	1
R_3	Open	2.567	2.187	2.157	1.143	1.258	
R_4	0	2.567	4.374	8.628	5.714	11.324	
C_1	0.5C	C	C	C	2C	2C	
R_1	0.407	0.407	0.378	0.378	0.173	0.141	
R_2	5.423	5.423	0.583	0.583	1.275	0.784	2
R_3	Open	Open	1.922	1.922	1.737	1.027	
R_4	0	0	1.922	1.922	8.687	9.244	
C_1	0.1C	0.1C	C	C	C	2C	

[a] Resistances in kilohms for a K parameter of 1.

Table 6-27. Second-Order Bessel (Constant-Time-Delay) Infinite-Gain MFB Filter Designs

Gain	Circuit Element Values[a]					
	1	2	4	6	8	10
R_1	1.466	1.100	0.841	0.984	0.731	0.770
R_2	1.466	2.199	3.363	5.903	5.850	7.697
R_3	1.919	1.919	2.510	1.430	2.887	2.194
C_1	0.3C	0.2C	0.1C	0.1C	0.05C	0.05C

[a] Resistances in kilohms for a K parameter of 1.

Table 6-28. Fourth-Order Bessel (Constant-Time-Delay) Cascaded MFB Filter Designs

Gain	Circuit Element Values[a]						Stage
	1	2	6	10	36	100	
R_1	0.692	0.519	0.539	0.415	0.431	0.370	
R_2	0.692	1.038	1.618	2.076	2.587	3.702	1
R_3	1.335	1.335	0.856	1.335	1.071	1.497	
C_1	0.3C	0.2C	0.2C	0.1C	0.1C	0.05C	
R_1	1.029	1.029	0.772	0.772	0.678	0.604	
R_2	1.029	1.029	1.543	1.543	4.065	6.036	2
R_3	1.429	1.429	1.429	1.429	1.085	1.218	
C_1	0.15C	0.15C	0.1C	0.1C	0.05C	0.03C	

[a] Resistances in kilohms for a K parameter of 1.

Table 6-29.   Third-Order Bessel
(Constant-Time-Delay)
Multiple-Feedback
Filter Design

	Circuit Element Values[a]
Gain	2
R_1	1.277
R_2	0.670
R_3	0.314
R_4	4.522

[a] Resistances in kilohms for a K parameter of 1.

Table 6-30.   Fourth-Order Bessel (Constant-Time-Delay)
Multiple-Feedback Filter Designs

	Circuit Element Values[a]				
Gain	2	4	6	8	10
R_1	0.152	0.135	0.126	0.119	0.114
R_2	1.236	1.583	1.863	2.111	2.338
R_3	0.458	0.207	0.154	0.127	0.109
R_4	0.355	0.692	0.847	0.957	1.044
R_5	1.626	1.198	1.202	1.239	1.282
R_6	1.626	3.595	6.008	8.672	11.537

[a] Resistances in kilohms for a K parameter of 1.

6.10.   Summary of All-Pass Constant-Time-Delay Filter Design Procedure

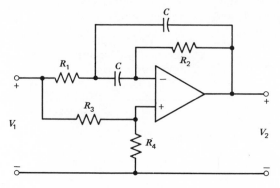

General circuit

Procedure

Given time-delay T_d, perform the following steps.

1.   Find f_0 in hertz for T_d in seconds from

$$f_0 = \frac{0.15915}{T_d}$$

or from the appropriate one of Fig. 6-9*a*, *b*, or *c*. (Note: Alternately, one could specify f_0 and find T_d by this procedure).

2.   Select a value of capacitance C and determine a K parameter from

$$K = \frac{100}{f_0 C'}$$

where C' is the value of C in microfarads. Alternately, K may be found from Fig. 6-8*a*, *b*, or *c*.

3.   Find the resistance values from Table 6-31 for $K = 1$, and multiply these values by the K parameter of step 2 to yield the resistances of the circuit.

4.   Select standard resistance values that are as close as possible to those indicated by the table and construct the filter in accordance with the general circuit.

Comments and Suggestions

Paragraphs (a), (b), and (c) of Sec. 6.7 apply directly.

The resistances R_3 and R_4 are determined in the design table to minimize the dc offset voltage. Other values may be used as long as the ratio $R_3/R_4 = 3$.

A specific example was given in Sec. 6.6

Table 6-31.   Second-Order All-Pass, Constant-Time-Delay Filter Designs

Circuit Element Values[a]	
R_1	0.398
R_2	0.531
R_3	0.838
R_4	0.279

[a] Resistances in kilohms for a K parameter of 1, gain is 0.25.

References

1. **J. L. Hilburn** and **D. E. Johnson,** *Manual of Active Filter Design,* McGraw-Hill, New York, 1973.

2. **J. G. Graeme, G. E. Tobey,** and **L. P. Huelsman** (eds.), *Operational Amplifiers: Design and Applications,* McGraw-Hill, New York, 1971.

3. **S. K. Mitra** (ed.) *Active Inductorless Filters,* IEEE Press, New York, 1971.

4. **S. K. Mitra,** *Analysis and Synthesis of Linear Active Networks,* John Wiley, New York, 1969.

5. **S. S. Haykin,** *Active Network Theory,* Addison-Wesley, Reading, Mass., 1970.

6. **L. P. Huelsman,** *Theory and Design of Active RC Circuits,* McGraw-Hill, New York, 1968.

7. **L. P. Huelsman,** *Active Filters: Lumped, Distributed, Integrated, Digital, and Parametric,* McGraw-Hill, New York, 1970.

8. **K. L. Su,** *Active Network Synthesis,* McGraw-Hill, New York, 1965.

9. *Linear Integrated Circuit D.A.T.A. Book,* Derivation and Tabulation Associates, Orange, N.J., Spring 1974.

10. *RCA Linear Integrated Circuits,* RCA Solid State Division, Somerville, N.J., 1970.

11. **R. Melen and H. Garland,** *Understanding IC Operational Amplifiers,* Howard W. Sams, Indianapolis, 1971.

12. *Radio Electronics* (monthly publication), Gernsback Publications, New York.

13. **F. C. Fitchen,** *Electronic Integrated Circuits and Systems,* Van Nostrand Reinhold, New York, 1970.

14. **A. B. Grebene,** *Analog Integrated Circuit Design,* Van Nostrand Reinhold, New York, 1972.

15. *Popular Electronics* (monthly periodical), Ziff-Davis, New York.

16. *Electronics* (biweekly periodical), McGraw-Hill, New York.

17. G. S. Moschytz and R. W. Wyndrum Jr., "Applying the operational amplifier," *Electronics,* December 9, 1968, pp. 98–106.

18. *Electronics Circuit Designers Casebook*, Electronics, New York, 1972.

19. E. A. Guillemin, *Synthesis of Passive Networks*, John Wiley, New York, 1957.

20. N. Balabanian, *Network Synthesis*, Prentice-Hall, Englewood Cliffs, N.J., 1958.

21. M. E. Van Valkenburg, *Introduction to Modern Network Synthesis*, John Wiley, New York, 1960.

22. L. Weinberg, *Network Analysis and Synthesis*, McGraw-Hill, New York, 1962.

23. G. C. Temes and S. K. Mitra, *Modern Filter Theory and Design*, John Wiley, New York, 1973.

24. A. Budak, *Passive and Active Network Analysis and Synthesis*, Houghton Mifflin, Boston, 1974.

25. A. Papoulis, "On the approximation problem in filter design," *IRE National Convention Record*, vol. 5, pt. 2, pp. 175–185, 1957.

26. R. P. Sallen and E. L. Key, "A practical method of designing RC active filters," *IRE Transactions on Circuit Theory*, CT-2, pp. 74–85, March 1955.

27. J. Tow, "A step-by-step active filter design," *IEEE Spectrum*, pp. 64–68, December 1969.

28. *Handbook of Operational Amplifier Active RC Networks*, Burr-Brown Research Corporation, Tucson, 1966.

29. J. Tow, "Design formulas for active RC filters using operational amplifier biquad," *Electron. Letters*, pp. 339–341, July 24, 1969.

30. L. C. Thomas, "The Biquad: Part I—Some practical design considerations," *IEEE Trans. Circuit Theory*, vol. CT-18, pp. 350–357, May 1971.

31. J. L. Hilburn, D. E. Johnson, and A. Eskandar, "Low-pass and high-pass designs for higher-order active filters," *Proc. 1974 IEEE Region 3 Conference and Exhibit*, April 1974.

32. G. Szentirmai, "Synthesis of multiple-feedback active filters," *BSTJ*, pp. 527–555, April 1973.

33. P. E. Fleischer and J. Tow, "Design formulas for biquad active filters using three operational amplifiers," *Proc. IEEE*, vol. 61, no. 5, pp. 662–663, May 1973.

34. P. R. Geffe, *Simplified Modern Filter Design*, Hayden, New York, 1963.

35. W. J. Kerwin and L. P. Huelsman, "The design of high performance active RC band-pass filters," *IEEE International Convention Record*, vol. 14, pt. 10, pp. 74–80, 1960.

36. L. P. Huelsman, *Theory and Design of Active RC Circuits*, McGraw-Hill, New York, 1968.

37. R. Brandt, "Active resonators save steps in designing active filters," *Electronics* April 24, 1972, pp. 106–110.

38. J. L. Hilburn and D. E. Johnson, "A fourth-order bandpass filter," *Proc. 17th Midwest Symposium on Circuits and Systems*, September 1974.

39. D. E. Johnson, J. L. Hilburn, and F. H. Irons, "Multiple-feedback higher-order active filters," *Proc. 1974 IEEE Region 3 Conference and Exhibit*, April 1974.

40. G. Hurtig III, "The primary resonator block technique of filter synthesis," *Proc. International Filter Symposium*, p. 84, April 1972.

41. R. M. Inigo, "Active filter realization using finite-gain voltage amplifiers," *IEEE Transactions on Circuit Theory*, vol. CT-17, pp. 445–448, Aug. 1970.

42.   **T. Deliyannis,** "RC active allpass sections," *Electron Letters*, vol. 5, pp. 59–60, February 1969.

43.   **L. Storch,** "Synthesis of constant-time-delay ladder networks using Bessel polynomials," *Proc. of the IRE*, vol. 42, no. 11, pp. 1666–1675, November 1954.

44.   **A. H. Marshak, D. E. Johnson,** and **J. R. Johnson,** "A Bessel rational filter," *IEEE Transactions on Circuits and Systems*, vol. CAS-22, November 1974.

Appendix A

Coefficients of Denominator Polynomial

$$s^n + b_{n-1}s^{n-1} + \cdots + b_2 s^2 + b_1 s + b_0$$

of Low-Pass Butterworth and Chebyshev Transfer Functions,
Normalized to a Cutoff Frequency of 1 rad/sec

Table A-1. Butterworth Filter: $s^n + b_{n-1}s^{n-1} + \cdots + b_1 s + b_0$

n	b_0	b_1	b_2	b_3	b_4	b_5	b_6	b_7
1	1.00000							
2	1.00000	1.41421						
3	1.00000	2.00000	2.00000					
4	1.00000	2.61313	3.41421	2.61313				
5	1.00000	3.23607	5.23607	5.23607	3.23607			
6	1.00000	3.86370	7.46410	9.14162	7.46410	3.86370		
7	1.00000	4.49396	10.09783	14.59179	14.59179	10.09783	4.49396	
8	1.00000	5.12583	13.13707	21.84615	25.68836	21.84615	13.13707	5.12583

Table A-2. 0.1 dB Chebyshev Filter ($\epsilon = 0.15262$):
$$s^n + b_{n-1}s^{n-1} + \cdots + b_1 s + b_0$$

n	b_0	b_1	b_2	b_3	b_4	b_5	b_6	b_7
1	6.55222							
2	3.31329	2.37209						
3	1.63809	2.62953	1.93883					
4	0.82851	2.02550	2.62680	1.80377				
5	0.40951	1.43556	2.39696	2.77071	1.74396			
6	0.20713	0.90176	2.04784	2.77908	2.96575	1.71217		
7	0.10238	0.56179	1.48293	2.70514	3.16925	3.18350	1.69322	
8	0.05179	0.32645	1.06667	2.15932	3.41855	3.56485	3.41297	1.68104

Table A-3.   0.5 dB Chebyshev Filter ($\epsilon = 0.34931$):
$$s^n + b_{n-1}s^{n-1} + \cdots + b_1 s + b_0$$

n	b_0	b_1	b_2	b_3	b_4	b_5	b_6	b_7
1	2.86278							
2	1.51620	1.42562						
3	0.71569	1.53490	1.25291					
4	0.37905	1.02546	1.71687	1.19739				
5	0.17892	0.75252	1.30957	1.93737	1.17249			
6	0.09476	0.43237	1.17186	1.58976	2.17184	1.15918		
7	0.04473	0.28207	0.75565	1.64790	1.86941	2.41265	1.15122	
8	0.02369	0.15254	0.57356	1.14859	2.18402	2.14922	2.65675	1.14608

Table A-4.   1 dB Chebyshev Filter ($\epsilon = 0.50885$):
$$s^n + b_{n-1}s^{n-1} + \cdots + b_1 s + b_0$$

n	b_0	b_1	b_2	b_3	b_4	b_5	b_6	b_7
1	1.96523							
2	1.10251	1.09773						
3	0.49131	1.23841	0.98834					
4	0.27563	0.74262	1.45392	0.95281				
5	0.12283	0.58053	0.97440	1.68882	0.93682			
6	0.06891	0.30708	0.93935	1.20214	1.93083	0.92825		
7	0.03071	0.21367	0.54862	1.35754	1.42879	2.17608	0.92312	
8	0.01723	0.10723	0.44783	0.84682	1.83690	1.65516	2.42303	0.91981

Table A-5. 2 dB Chebyshev Filter ($\epsilon = 0.76478$):
$$s^n + b_{n-1}s^{n-1} + \cdots + b_1s + b_0$$

n	b_0	b_1	b_2	b_3	b_4	b_5	b_6	b_7
1	1.30756							
2	0.82325	0.80382						
3	0.32689	1.02219	0.73782					
4	0.20577	0.51680	1.25648	0.71622				
5	0.08172	0.45935	0.69348	1.49954	1.70646			
6	0.05144	0.21027	0.77146	0.86701	1.74586	0.70123		
7	0.02042	0.16609	0.38251	1.14444	1.03922	1.99353	0.69789	
8	0.01286	0.07294	0.35870	0.59822	1.57958	1.21171	2.24225	0.69606

Table A-6. 3 dB Chebyshev Filter ($\epsilon = 0.99763$):
$$s^n + b_{n-1}s^{n-1} + \cdots + b_1s + b_0$$

n	b_0	b_1	b_2	b_3	b_4	b_5	b_6	b_7
1	1.00238							
2	0.70795	0.64490						
3	0.25059	0.92835	0.59724					
4	0.17699	0.40477	1.16912	0.58158				
5	0.06264	0.40794	0.54886	1.41498	0.57443			
6	0.04425	0.16343	0.69910	0.69061	1.66285	0.57070		
7	0.01566	0.14615	0.30002	1.05184	0.83144	1.91155	0.56842	
8	0.01106	0.05648	0.32076	0.47190	1.46670	0.97195	2.16071	0.56695

Appendix B

Second-Order Factors of Denominator Polynomial

$$\prod_{i=1}^{n} (s^2 + a_i s + b_i)$$

of Low-Pass Butterworth and Chebyshev Filters,
Normalized to a Cutoff Frequency of 1 rad/sec

Table B-1. Butterworth Filter: $\prod_{i=1}^{n} (s^2 + a_i s + b_i)$

n	a_1	b_1	a_2	b_2	a_3	b_3	a_4	b_4
1	1.41421	1.00000						
2	0.76537	1.00000	1.84776	1.00000				
3	1.51764	1.00000	1.41421	1.00000	1.93185	1.00000		
4	0.39018	1.00000	1.11114	1.00000	1.66294	1.00000	1.96157	1.00000

Table B-2. 0.1 dB Chebyshev Filter: $\prod_{i=1}^{n} (s^2 + a_i s + b_i)$

n	a_1	b_1	a_2	b_2	a_3	b_3	a_4	b_4
1	2.37209	3.31329						
2	0.52827	1.32981	1.27536	0.62282				
3	0.22940	1.12953	0.62674	0.69646	0.85614	0.26339		
4	0.12797	1.06964	0.36443	0.79901	0.54540	0.41627	0.64334	0.14563

Table B-3. 0.5 dB Chebyshev Filter: $\prod\limits_{i=1}^{n} (s^2 + a_i s + b_i)$

n	a_1	b_1	a_2	b_2	a_3	b_3	a_4	b_4
1	1.42562	1.51620						
2	0.35071	1.06352	0.84668	0.35641				
3	0.15530	1.02302	0.42429	0.59001	0.57959	0.15700		
4	0.08724	1.01193	0.24844	0.74133	0.37182	0.35865	0.43859	0.08805

Table B-4. 1 dB Chebyshev Filter: $\prod\limits_{i=1}^{n} (s^2 + a_i s + b_i)$

n	a_1	b_1	a_2	b_2	a_3	b_3	a_4	b_4
1	1.09773	1.10251						
2	0.27907	0.98650	0.67374	0.27940				
3	0.12436	0.99073	0.33976	0.55772	0.46413	0.12471		
4	0.07002	0.99414	0.19939	0.72354	0.29841	0.34086	0.35200	0.07026

Table B-5. 2 dB Chebyshev Filter: $\prod\limits_{i=1}^{n} (s^2 + a_i s + b_i)$

n	a_1	b_1	a_2	b_2	a_3	b_3	a_4	b_4
1	0.80382	0.82325						
2	0.20977	0.92868	0.50644	0.22157				
3	0.09395	0.96595	0.25667	0.53294	0.35061	0.09993		
4	0.05298	0.98038	0.15089	0.70978	0.22582	0.32710	0.26637	0.05650

Table B-6. 3 dB Chebyshev Filter: $\prod\limits_{i=1}^{n} (s^2 + a_i s + b_i)$

n	a_1	b_1	a_2	b_2	a_3	b_3	a_4	b_4
1	0.64490	0.70795						
2	0.17034	0.90309	0.41124	0.19598				
3	0.07646	0.95483	0.20890	0.52182	0.28535	0.08880		
4	0.04316	0.97417	0.12290	0.70358	0.18393	0.32089	0.21696	0.05029

Index